Lecture Notes in Economics and Mathematical Systems

Managing Editors: M. Beckmann and W. Krelle

364

Balazs Horvath

Are Policy Variables Exogenous?

The Econometric Implications of Learning
while Maximizing

Springer-Verlag Berlin Heidelberg GmbH

Managing Editors
Prof. Dr. M. Beckmann
Brown University
Providence, RI 02912, USA

Prof. Dr. W. Krelle
Institut für Gesellschafts- und Wirtschaftswissenschaften
der Universität Bonn
Adenauerallee 24–42, D-5300 Bonn, FRG

Author
Balazs Horvath
International Monetary Fund
700 19th Street, N.W.
Washington, D.C. 20431, USA

KOPINT-DATORG
1389 Budapest
Dorottya u. 6, Hungary

ISBN 978-3-540-54287-2   ISBN 978-3-642-58211-0
DOI 10.1007/978-3-642-58211-0

2142/3140-543210 – Printed on acid-free paper

Acknowledgements

This work grew out of a line of research initiated by Marc Nerlove at the University of Pennsylvania. I thank him for the topic, his guidance as I struggled to put the pieces together and his financial support. I thank András Simon at the University of Economics, Budapest for his instrumental role in my getting the opportunity to study at the University of Pennsylvania. I am indebted to Lawrence Klein and Peter Pauly at Project LINK for having made my studies possible in my first two years at the University of Pennsylvania, provided financial support and an exciting working environment. The research underlying this work was partially supported by a grant from the National Science Foundation to the University of Pennsylvania (SES 8921715).

I benefitted most from the comments of my dissertation advisor, Marc Nerlove and Christian Gourieroux, who visited the University of Pennsylvania in the spring of 1990. I also gratefully acknowledge extremely helpful comments from T. W. Anderson, Viktoria Dalko, Javier Gardeazabal, Nicholas Kiefer, Richard Kihlstrom, Wilhelm Krelle, George Mailath, Roberto Mariano, Bruce Mizrach, Marta Regulez, Rafael Rob, Christopher Sims, Stefano Siviero, Douglas Willson, seminar participants at the University of Pennsylvania, Rutgers and Princeton Universities, the fifth FUR Conference at Duke University and at the 6th World Congress of the Econometric Society in Barcelona as well as expert computational advice from Tilda Horvath and George Theall. Of course, any remaining errors are mine.

I am grateful to my wife, Tilda and my daughters, Réka and Flóra for their love and because they believed in me even when I didn't. Finally, my gratitude is expressed to my parents.

Abstract
Are Policy Variables Exogenous?
The Econometric Implications of Learning while Maximizing

Balazs Horvath

        This study explores the econometric implications of learning by
economic agents. A distinction between active and passive learning is made.
On the basis of an argument on the curvature of the value function arising
in a dynamic programming approach to the general formulation of the problem,
active learning is shown to be the rule rather than the exception. To
provide a framework for the analysis, a paradigmatic model is presented in
which the government maximizes the discounted sum of tax revenues subject to
the constraint imposed by a Laffer curve involving a parameter initially not
precisely known but about which Bayesian learning occurs. The paradigmatic
model is nontrivial and dynamic by virtue of the presence of learning about
the unknown parameter. The government strikes an optimal balance between
maximization of current payoff and generation of future information which
enhances the efficiency of maximization in subsequent periods.
        The issue of exogeneity of policy variables is addressed. It is
demonstrated that learning affects the exogeneity status of policy variables
and has implications analogous to the phenomenon in the focus of the Lucas
critique. An additional constraint to augment the exogeneity definitions in
Engle, Hendry and Richard (1983) is proposed.
        A simulation exercise based on the model presented supplies
additional insights and quantitative evidence on the structure of the
problem. Active learning is proven to be a distinct cause of time
inconsistency of optimal plans, the extent of which is quantified for some
specific examples.
        It is argued that learning is not dismissable as a merely
transitory source of time inconsistency and loss of parameter invariance.
        Finally, the data generated are used to perform empirical
exogeneity tests in the manner of Granger (1980) and Sims (1972), utilizing
results from the survey of Geweke (1984). The results support the claim that
the effects of Bayesian learning can be empirically detectable.

Table of Contents

Chapter IV

    Simulation

Chapter V

    Tests for Exogeneity

Chapter VI

    Summary, Directions for Future Research

Appendices

List of Tables

List of Illustrations

I. Introduction

I.1 Motivation and Definition of Topic

To provide motivation and to help define the topic of this study, important links between specific areas of economic theory are first highlighted.

(i)   Learning and Rational Expectations Theory

In a standard rational expectations setting, agents in equilibrium have all the information about the model that enables them to correctly forecast future payoff-relevant variables. What rational expectations theory in its standard form does not tell us is what happens outside a rational expectations equilibrium. Less than complete knowledge of the model is a possible way to represent a situation outside the rational expectations equilibrium. It is natural to assume that agents recognize error and optimally utilize all available external information to improve on their information level, i.e. learn. Based on the information acquired by learning they modify their behavior. Under certain conditions learning steers the economy to the rational expectations equilibrium (Spear (1989), Blume, Bray and Easley (1982), Townsend (1983)). This literature shows that learning is a possible mechanism to acquire the necessary level of information that agents are assumed to possess in a rational expectations equilibrium and hence there is a clear link between rational expectations theory and the theory of learning. This fact is also emphasized among others by Friedman (1975), Pesaran (1987) and DeCanio (1979).

(ii)   Rational Expectations and Econometrics

The equilibrium consequences of the rational expectations hypothesis are discussed in a considerable body of literature - cf. Wallis (1980), Hansen and Sargent (1980), Pesaran (1987, Chapter 6). Another equilibrium implication has been forcefully described by Lucas in the well known Lucas (1976) critique of conditional policy experiments based on an estimated econometric model leading to loss of much of the appeal of traditional macroeconometric models. The essence of the Lucas criticism has been summarized among others in Sargent (1981), (1987) from a time series analysis point of view and in Engle, Hendry and Richard (1983) from an econometric point of view. In the latter, theoretical econometric formulation, the Lucas critique is formulated as the possible failure of policy variables to be super exogenous (exact definitions of exogeneity concepts are found in Section III.1).

Sargent (1987) examines a standard task for an economic agent in a rational expectations framework: forming optimal forecasts of future values of a variable Y based on the history of that process and a related one, X. He points out that the way this can be achieved hinges on a characteristic of the joint process of the variables - the notion of Granger causality (also defined in Section III.1). This notion in turn is an element of Engle, Hendry and Richard's concept of strong exogeneity, and also plays a role in the empirical tests for various forms of exogeneity, as will be seen in this study. At this stage we note therefore that rational expectations theory has a bearing on econometrics in general and on the issue of exogeneity of policy variables in macroeconometric models in particular. Given that the concept of rational expectations has been shown to have numerous vitally important equilibrium implications for econometric practice, it is natural to raise the possibility of important out of equilibrium implications as well.

(iii)   Learning and Econometric Exogeneity of Policy Variables

Although the previous two links already imply an indirect relationship between learning and econometric exogeneity of policy variables, the case will be made that a direct relationship exists. Learning can be defined as an informational feedback: data generated by the environment contain information on the structure of the data generating process which is utilized by the agent to refine his information about the environment, in particular, about the constraint he faces when maximizing his payoff over some time horizon. In most cases the choice of variables controlled by the agent have an impact on the amount of information contained in subsequent data generated by the environment and observed by the agent. Since this impact is known to the agent, optimizing choice of policy variables will be influenced by earlier observations, i.e. a feedback is introduced.

The assumption that policy variables are exogenous is a routine one in empirical econometrics. In fact identification of equations involving these variables often hinges on the validity of this assumption. However, exogeneity as perceived in Koopmans (1950) and by econometricians following his approach is incompatible with the presence of feedbacks.

In-depth analysis of this area therefore necessitates finding answers to questions like:

How can learning be modelled in an intuitive and satisfactory way?

What forms of learning behavior can be distinguished?

Are policy variables still exogenous when learning is performed by the policymaker?

Precisely what, if any, aspects of exogeneity are affected?

Is the presence of learning detectable by an outside econometrician having access only to data on the policy variable and the outcome variable and possessing a reasonably specified model?

How does the presence of learning affect the validity of policy experiments using an econometric model and the time consistency of the policymaker?

The above questions can be classified into vastly different areas such as economics of information, econometrics, decision theory, optimal control and macroeconomic policy analysis. Yet they constitute a well defined and meaningful topic - the set of issues to be addressed in this study.

The approach advocated in Kiefer (1988-89) is adopted in that the study does not consider general equilibrium models and in general, disregards interactions between agents. The justification is that the optimizing behavior of agents in information gathering first needs to be analysed before more general models can be formulated in the framework of a general equilibrium model. The focus, as the title suggests, will be on the exogeneity of policy variables. Of course, the world is a system of interdependent variables. Hence, by assuming variables to be exogenous, one always commits some error. The question is one of magnitudes: how important feedbacks to the exogenous variables are. The argument presented in this work is that once all the pieces are put together, the feedback introduced by learning cannot generally be ignored. It is significant enough to make assuming policy variables to be exogenous in a learning environment a serious misspecification. Inasmuch as learning is accepted to be the out of equilibrium driving force guiding economic systems to rational expectations equilibria, we have identified an out of equilibrium implication of the rational expectations hypothesis.

I.2 Overview and Literature Survey

The focus of this study, as broadly described above, is on the behavior of a rational agent with less than perfect information. The agent's information set consists of prior information plus sufficient statistics of all the observations containing information relevant for the calculation of a fully optimal solution to the agent's problem. He refines this information set by means of learning using observations generated by the (partially unknown) environment.

The environment is assumed to be a nonchanging data generating process. This does not mean that coefficients in the model are necessarily constant over time, but it does mean that if they change, they do so according to rules involving constant parameters that are included in the overall description of the data generating process.

A rational agent performing learning will seek to utilize all available information in an optimal way. In subsequent chapters, learning is formalized by repeated application of Bayes' rule, shown formally to be an optimal information processing mechanism in a rather general setting in Zellner (1988). Bayes' rule is widely utilized to model learning in the literature (e.g. by Easley and Kiefer (1988), DeGroot (1970), Grossman, Kihlstrom and Mirman (1977), Crawford (1973) among others).

This form of learning is in the class of learning mechanisms referred to in the literature as rational learning - cf. Pesaran (1987, p.34), Blume, Bray and Easley (1982), Bray and Kreps (1986), Spear (1989). The concept of rational learning involves the assumption that the specification of the true structural model is known to the agent but the value of some parameters is not. Then the agent refines his initial beliefs about the unknown parameters as new information becomes available. For the agent to be rational, this refinement procedure must be optimal in some sense. As already mentioned above, Bayesian updating of belief distributions is such.

It is noted in Bray and Kreps (1986) that assuming Savage-rational agents (i.e. ones performing learning about the values of parameters in the correctly specified model by Bayesian updating of subjective belief distributions) is only one step towards the analysis of the fundamental question: from where do agents obtain the information that they are supposed to have in a rational expectations equilibrium? In fact, since a number of additional assumptions have to be made to ensure that beliefs converge to the true values of the initially unknown parameters, the fundamental question in models of rational learning is merely pushed one stage back. Given the emphasis of this study on econometric implications however, utilizing this setup is warranted if this one step in itself is sufficient for interesting econometric results to emerge.

A classical, rather than Bayesian treatment of essentially the same problem is given in the vast literature on adaptive control initiated by Bellman (1961). It does not require the agent to hold a well specified prior belief or possess strong computational abilities but it cannot lay claim to optimal information processing as Bayes rule can. Typically a linear-quadratic optimal control problem is posed. Major references include Chow (1981), Kendrick (1982), Tse (1974), Bar-Shalom and Tse (1976), Hughes-Hallett and Rees (1983), Marquez and Pauly (1986), Holly and Hughes-Hallett (1989). Because the subject matter of this study is in the realm of optimal control theory as well, the term control variable will often be used for the policy variable.

In contrast to rational learning, boundedly rational learning is less demanding in its assumptions on the amount of information and the computing capabilities of the agent. As Pesaran (1987) points out, postulating boundedly rational learning amounts to assuming that agents know the true reduced form of the model generating the observations - a requirement generally viewed as more realistic - cf. Marcet and Sargent (1987a), (1987b). Work in both areas is also geared towards developing a justification for the rational expectations hypothesis by providing out of equilibrium information gathering algorithms which imply that the system converges to a rational expectations equilibrium. This goal is known in the literature as proving the stability of the rational expectations equilibrium - cf. Cyert and DeGroot (1974), Fourgeaud, Gourieroux and Pradel (1986), Bray and Savin (1986). If the rational expectations equilibrium is attained in the limit, then learning is complete, otherwise it is incomplete. For an example when this convergence does not occur and the profound policy implications, consult Mizrach (1989).

Incomplete learning does not imply lack of rationality, however. First, it may not pay to learn completely, even if it is a possibility - a point repeatedly emphasized in Kiefer (1988-89). Second, as Rust (1988-89) points out, in models where identification conditions are not met in the sense that a different value of the unknown parameter: θ' implies the same optimal controls as the true value θ^*, a learning mechanism that has beliefs converging to θ' "works": it generates decision rules of the agent converging to a decision rule that is optimal under full information (i.e. knowing θ^*). Thus it also generates optimal behavior despite the fact that learning is incomplete.

As the above example illustrates, the issue of identifiability arises in the context of learning. One sense in which identification can be lost has been described above. There is another sense in which this problem can arise: if there are too many aspects of the model which can be learned about. Consider the case when the agent learns about the constraint he faces as well as about his objective function[1]. (Notes appear at the end of each chapter.) Let $C(\theta_1)$ and $W(\theta_2)$ denote the constraint and the objective function respectively, each a function of a distinct set of parameters not precisely known initially. Index these by the beliefs held by the agent about the unknown parameter vectors: e.g. $C(\theta_1)_i$ denotes the perception of the constraint if belief i is held by the agent about θ_1. Then clearly, it may be the case that both $\left\{ \begin{array}{c} C(\theta_1)_i \\ W(\theta_2)_i \end{array} \right\}$ and $\left\{ \begin{array}{c} C(\theta_1)_j \\ W(\theta_2)_j \end{array} \right\}$, $i \neq j$ are compatible with the observed behavior of the system for some finite number of observations even if θ_1 is individually identified for θ_2 known and vice versa. Thus these two constellations are observationally equivalent, i.e. θ_1 and θ_2 are not identified. It follows that the assumption that parameters are identifiable using the flow of data available to the learning agent must be made explicit - a point emphasized in Bray and Kreps (1986, chapter V).

While it is true that imposing additional structure on the various aspects of learning is a possible way to achieve identification, this is not the approach utilized in the subsequent chapters. Instead, it will be assumed that the agent knows every aspect of the model except the value of one of the parameters in the constraint. An illustrative multiperiod model will be presented with an initially less than perfectly informed policy maker, who is learning about the structure of the environment at the same time he is maximizing his social welfare function subject to the constraint posed by it.

To be more specific, consider a policymaker whose social welfare function includes only expenditures on public goods and who must balance his budget. Thus he maximizes the discounted sum of tax revenues by determining, in each period, a tax rate (or schedule of rates) to be applied to individuals in the economy. The constraint the policy maker is assumed to face is the so-called Laffer curve - to be precisely specified shortly. It will be assumed that this tax revenue function has a unique maximum somewhere between a tax rate of 0% and a rate of 100%, that the function is subject to some stochastic variation and that the parameter determining the location of the maximum is imperfectly known. At each point of time, however, the policymaker has a prior on the parameter, conditional on observations up to that point.

The problem can also be described as one of optimally controlling a stochastic process with an unknown parameter. The agent controlling the process has beliefs about these parameters embodied in a (prior) probability distribution. He is assumed to refine his beliefs - or learn - as time goes on in the sense of updating his prior distribution via Bayes' rule, using the information that emerges as the process proceeds. With more information, more efficient optimization is possible in subsequent periods. Therefore, the agent in any given period has to make an optimal trade-off between two competing goals: maximizing current payoff given his current level of information versus maximizing the expected information yield about the unknown parameters. Given a prior that does not rule out prospects for generation of valuable information (cf. Pesaran, 1987 and also section II.4), this trade-off is always present in multiperiod problems of this sort. It is in this sense that this problem is referred to in the optimal control literature as a dual control problem.

The distinction between open loop policy, passive and active learning is to be made. Open loop policy is one that is based on a nonchanging information set. Note that this concept encompasses both the full information case and the case when information is less than perfect but is not augmented, i.e. no learning occurs. Passive learning stands for incorporating any information that happens to be generated as a result of optimizing behavior ignoring experimental design aspects of the problem. Active learning occurs when the policy maker optimally trades off current payoff for future information expected to be generated, i.e. he optimally mixes open loop control and experimental design.

The Laffer curve is obtained as a reduced form arising from the interaction of two distinct tax effects: first, the effect on total output and second, the effect on the amount of evasion or of output transferred to the underground sector. The derivation for the additive case is found in Appendix A, which also provides the rationale for the choice of functions in chapter II.

Utilizing the Laffer curve example has some advantages and disadvantages. Clearly, it is a crude approximation of reality only, which disregards important aspects of the game between the government and the public as well as the way tax income is spent by the government. No real world policymaker would optimize in the simple minded manner assumed in this study, disregarding the effect of taxation on the price level and relative prices or on international competitiveness. The effect of the tax burden on output - and by implication, on employment - is however accounted for. The emphasis is not on the exact interpretation: the model presented shortly can also be interpreted as one describing a monopolist maximizing the discounted net revenue subject to an unknown demand curve. While such an interpretation may appear more plausible - indeed, it is the interpretation most often encountered in the literature, e.g. in Grossman, Kihlstrom, and Mirman (1977) and Kiefer (1988-89), it has its own set of drawbacks. First, to allow for the theoretically interesting case of active learning, a small number of players must be assumed - a natural environment for strategic behavior. The alternative to current payoff for the monopolist is information, a public good - a point raised in a different context in Jovanovic and Lach (1989) and Rob (1988). Thus, when the monopolist trades off current reward for information, a free-rider problem may arise: potential entrants can observe the price and obtain information on demand "for free". In the formulation of chapter II the policymaker holds an uncontestable monopoly of taxation and thus the information generated by any experimentation cannot be used by any other agent. Also, the atomistic nature of the public (the potential other player) makes it improbable that a stable coalition can be formed to "cheat". (For a related argument see Grunberg and Modigliani (1954).) To summarize, this setting avoids complications that arise from strategic behavior and from the public good nature of information, the good for which current payoff is traded off. These are hard to plausibly rule out in other contexts and would complicate the analysis without necessarily yielding much additional insight. The absence of strategic behavior will be essential in proving that learning implies via a distinctly different channel the same phenomenon as that in the focus of the Lucas critique. It is worth mentioning at this point that the inherent game aspects of the monopolist example have been addressed in Mirman, Samuelson and Urbano (1989) and Mirman, Samuelson and Schlee (1990).

The policymaker's problem is formulated as a simple stochastic multiperiod optimization problem which, when cast in a dynamic programming framework, yields a value function. To make room for learning, initially some quantities must be less than precisely known. To make learning nontrivial, noise or variability must enter the system at least in one place. Once learning is present, it generates a link between present period's actions and future beliefs, that is, it introduces dynamics into an originally static problem.

In chapter II, a model is specified that is argued to capture the salient features of the problem. Its well defined nature and its properties are established. Then the condition for the occurrence of active learning is sought. Basically it is found to concern the curvature of the value function in beliefs as developed in that chapter: it must be convex. As will be seen, a twist is given to the problem by the fact that the current choice of the control variable (in our case r_t, the tax rate) affects beliefs thereafter. Whether it pays to actively learn, is also affected by the discount factor δ. If it is too low, the policymaker discounts future expected gains too much to be able to recoup the portion of payoff foregone in the present period. Distributional assumptions can also rule out active learning, as formally shown in section II.4.

There is thus an interaction between the local curvature of the value function in beliefs, the choice of r_t and the value of δ in determining the optimality of active learning. Additional complications include the length and possible endogeneity of the time horizon, different assumptions about risk aversion and an intrinsically dynamic state evolution equation.

To proceed, the value function in its modified form (as a function of the belief distribution) is shown to be always convex (though not necessarily strictly convex in any period, which would make active learning strictly optimal). A value function that is affine in each period in beliefs is then argued to be the exception, rather than the rule. Consequently, at least for the class of models considered, the importance of active learning should be more than negligible. This theoretical result contrasts with the unanimous empirical finding to the contrary in the literature utilizing linear-quadratic optimal control (e.g. Kendrick (1982), Marquez and Pauly (1986)).

Having formalized the problem faced by the policymaker in a specific way, Chapter III explores the implications for econometric practice. Building mainly on work summarized in Engle, Hendry and Richard (1983), the answer is given to the question posed in the title: can policy variables that are generated by an agent performing learning in the sense described in chapter II be regarded as exogenous in econometric work? The answer is no for each possible goal of the econometrician utilizing the data: inference on (identified) parameters of interest, prediction and conditional policy experiments. Elaborating on a point made originally in Hughes-Hallett and Rees (1983), learning is shown to imply the same loss of parameter invariance to changes in policy regimes as the one addressed by the Lucas critique, via a distinctly different channel. Finally, the data generating process in an environment with a learning agent is shown to be nonstationary.

Chapter IV describes the results of a simulation exercise based on the specification derived in chapter II. Passive and active learning are contrasted, the sensitivity of results to elements in the specification is assessed. Finally, building on an idea in Hughes-Hallett and Rees (1983) as with the Lucas critique, (active) learning is argued to be a distinct possible cause of time inconsistency of optimal plans in a multiperiod optimization problem.

Chapter V concludes by applying a class of standard exogeneity tests to data generated in a learning environment. Several variants of exogeneity tests are applied to data generated in the simulations under different assumptions on the mode of learning and different sets of values for the parameters in the specification.

Appendix A describes the Laffer curve and works out its additive form under the assumption that noise enters the system at the observation stage only. It also contains a general version of the condition for active learning utilized in chapter II. Appendix B discusses a result encountered in the existing literature in which in the initial stages of active learning, controls will be of greater magnitude than later, since larger values generate more information. This result is not a general one, but holds for the specification of chapter II and will be referred to as the "bigger is better" result. Appendix C contains proof of the Lemma utilized in chapter II. Appendix D derives an expression for a partial derivative that plays an important role in the condition for active learning, and shows that it is positive. It also derives the condition for the policymaker's problem described in the study to be well defined in every respect.

Notes

1. Learning about the objective function is compatible with constant, well-defined preferences. An example is adjusting penalties corresponding to undesirable effects as these become worse than expected - cf. the description of instrument instability in Hughes-Hallett and Rees (1983, p.121). For an interesting discontinuity property proving that instrument instability is potentially a serious concern and it cannot be simply dismissed in many cases, consult Sims (1974).

II. A Paradigmatic Example

II.1 The Model: Tax Rate Determination under Simultaneous
Optimization and Learning

The theoretical foundations for this class of models have been summarized in Easley and Kiefer (1988). A discrete time decision problem is considered where the decisionmaker chooses an action r in each period to maximize total expected discounted reward depending on the action chosen and the outcome, a random variable. The conditional distribution $f(.|r, \beta)$ of the outcome given the action depends on an initially unknown parameter β. The decisionmaker begins with a prior belief about the unknown parameter and at the end of each period updates it via Bayes' rule utilizing the latest observations on the action taken and the outcome. Easley and Kiefer take the additional simplifying step of integrating out the outcome and redefining the maximand to be the total expected discounted mean reward where the mean is calculated with respect to the conditional distribution $f(.|r, \beta)$ and the belief distribution.

In our context the decision is made by a policymaker choosing the tax rate r. The outcome coincides with the reward: current tax revenue R as determined by a Laffer curve plus an additive error, therefore the above simplification is "built-in". The time horizon T is finite and known. The maximand corresponding to Easley and Kiefer's redefined maximand will be given by (6) below.

The underlying assumptions are reviewed first. They follow in spirit the set of assumptions made by Easley and Kiefer, and a comparison between the two will follow. The specification of the model introduced thereafter will comply with the assumptions presented below.

Assumptions

(i)   the action space is [0,1] - an unchanging compact, convex subset of the real line R^1

(ii)   the observation space is $[0,\bar{R}]$ - a compact subset of the real line R^1, where $\bar{R}$ is the maximum of the Laffer curve, unique and finite by assumption

(iii)   the parameter space is the real line R^1

(iv)   the conditional distribution $f(.|r, \beta)$ is continuous in r and β

(v)   the reward function, given by the Laffer curve is continuous in r and β

(vi)   the expected reward has a single maximum in r for any belief

(vii)   the discount factor $\delta \in (0,1]$

(viii) the support of the distribution representing beliefs about β is the real line R^1

(ix)   the belief distribution is conjugate

(x)   the time horizon is finite

With the exception of (iii), (vii), (viii), (ix) and (x), these are implied by the corresponding assumptions in Easley and Kiefer (1988, section II). Assumption (ix) is made only to ensure that Bayesian recursions are simple to handle. Assumption (x) is a departure: our focus is not on asymptotic results. Assumptions (iii), (vii) and (viii) are more general: the discount factor can be equal to unity because of the finite time horizon considered. Letting the belief distribution have infinite support facilitates the use of the normal distribution for representing beliefs, but comes with a price of having to impose additional structure on the problem - details are spelled out in Appendix D.

The Laffer curve specified below will be in a multiplicative form. The reason for considering this instead of the additive Laffer curve of Appendix A is the following. Beliefs evolve according to the intertemporal update rules provided by Bayes' rule. They however must in all periods comply with the Laffer restrictions, namely that tax rates of 0 or 1 always result in 0 revenue. In the additive version the Laffer restrictions imply a deterministic restriction on the set of parameters the policymaker is learning about. If there is only one parameter however, these two requirements are generally in conflict. One way out is to increase the number of parameters to 2 and impose the Laffer restriction as an identifying restriction: this reduces the dimension of the parameter space to 1 again. An alternative is to apply the approach utilized here which consists of factoring the objective function in a way that a portion of it involves no unknown parameters and delivers the properties required at the same time as having a plausible interpretation. The rest of the maximand contains the parameter about which learning occurs. This approach of factoring the constraint works in general if the restrictions to be incorporated can be formulated as zero restrictions. Yet another alternative could be to argue that the restrictions are nonstochastic constraints that can be directly incorporated into the objective function via a constant Lagrange multiplier and then optimization is to be carried out with respect to this augmented objective function. This approach is utilized in MacRae (1972) in a different context.

To proceed, let the Laffer curve be given by

$$R(\tau) = \tau \, Q(\tau) \, S(\tau) \qquad\qquad (1)$$

$$R(0) = R(1) = 0 \qquad\qquad (2)$$

where τ is marginal tax rate assumed to be same for the whole economy, so it is also the average tax rate; $Q(\tau)$ is total output and $S(\tau)$ is the evasion factor. Specify:

$$Q(\tau_t) = \bar{Q} \, (1 - \tau_t) \quad \text{and} \qquad\qquad (3)$$

$$S(\tau_t) = \alpha - \beta \, \tau_t + u_t \qquad\qquad (4)$$

where β is an unknown parameter, u_t is an i.i.d. doubly truncated random variable distributed as $N(0, s^2)$ and $\bar{Q}$, α and s^2 are known (positive) constants[1]. The reason for the double truncation will be explained shortly, it basically ensures that the policymaker does not come to hold extremely unreasonable beliefs due to an extreme sequence of realizations of the noise term. Without loss of generality we can assume $\bar{Q}=1$. The parameter β is the only unknown. We shall only be considering the case $\beta>0$. Note that u_t has a probability distribution symmetric around zero[2].

This specification enables us to achieve maximum simplicity and it imposes (2), the deterministic Laffer restrictions independently of the evolution of beliefs. Thus, learning with respect to the unknown parameter can proceed in an unrestricted manner. Passive learning corresponds to an approach treating periods separately. An actively learning agent on the other hand maximizes the total discounted sum of revenues, optimally trading off some of the obtainable current revenue for extra information generated. Hence the objective function for a passive learner is just $E_t R(r_t)$ in each period while that for an active learner is developed below. The precise definition of terms such as current information and E_t will also be given.

The next step is to formulate the policymaker's problem as a multiperiod problem with finite, known horizon T. For generality set $T \geq 2$. In period 1 the policymaker seeks to

$$\max_{\substack{\{r_t\}_{t=1}^{T} \\ r_t \in [0,1]}} E_1 \sum_{t=1}^{T} \delta^{t-1} R(r_t) \; . \tag{5}$$

where the discount factor $\delta \in (0,1]$ is a known constant. The policymaker chooses the tax rate r_t for each period so as to maximize (5) given the available information in the current period. The information set contains sufficient statistics of all payoff-relevant parameters. For those parameters that are assumed to be known, they are the values of these parameters themselves. For the sole unknown quantity, β, it is the probability distribution embodying current beliefs. Hence, the information set consists of the values of the known parameters and the current belief distribution about β updated via Bayes' rule utilizing all observations that have become available by the current period. The period objective function itself is static. There is a connection between time periods however, via the evolution of beliefs.

The policymaker's optimization problem, given by (5), may be written using (1), (3) and (4) equivalently as:

$$\max_{\substack{\{r_t\}_{t-1}^T \\ r_t \in [0,1]}} E_1 \sum_{t=1}^{T} \delta^{t-1} \left\{ r_t (1 - r_t) [\alpha - \beta r_t + u_t] \right\},$$

or to emphasize sequential reoptimization, as

$$\max_{\substack{\{r_t\}_{t-1}^T \\ r_t \in [0,1]}} E_1 \sum_{t=1}^{T} \delta^{t-1} E_{t-1} \left\{ r_t (1 - r_t) [\alpha - \beta r_t + u_t] \right\}. \qquad (6)$$

In interpreting the expectation above some care must be exercised. First, since the parameter β is unknown to the policymaker, the expectation will involve the current distribution embodying the beliefs held by the policymaker. It will also involve taking the current expectation of the noise term u_t: this expectation is always zero by assumption. Given our set of assumptions, these two will be orthogonal in the sense that the expectation operators in (6) simply become $E_t^{(u)} E_t^{(\beta)}$ where the superscript indicates the distribution with respect to which the expectation is to be computed and the subscript indicates the information set on which the expectation is conditioned. Note that the i.i.d. assumption on u_t is vital here.

To proceed, let us obtain a distribution that can be reasonably argued to embody the beliefs of the policymaker on the unknown parameter β. Define precision as the reciprocal of variance: $h - \sigma^{-2}$, and write $N(m, h)$ instead of the usual $N(m, \sigma^2)$ for a normal distribution with mean m and variance σ^2. In the policymaker's problem some probability weight may be placed on the parameter β being negative, this still leaves the problem well defined as long as m_t is strictly positive. Thus it is not unreasonable to assume that the prior probability density function for β is given by a normal density. Let $m_t = E_t(\beta)$ denote the mean belief in the t-th period.[3] Denote the prior by $p_1(\beta) = N(m_1, h_1)$ with $m_1 > 0$. Let

$$e_{t-1} = \beta \, r_{t-1} - u_{t-1}. \qquad (7)$$

To ensure that e_{t-1} can be treated as observable, it is assumed that the policymaker can observe R_{t-1}, i.e. the revenue generated in the previous period precisely. Then

$$e_{t-1} = \alpha - \frac{R_{t-1}}{r_{t-1}(1-r_{t-1})} \qquad (7')$$

is readily computable.

Assuming Bayesian updating we have the following update rules (Appendix B contains the derivation):

$$h_t = h_{t-1} + r^2_{t-1} \qquad (8)$$

$$m_t = \frac{m_{t-1}h_{t-1} + r_{t-1}e_{t-1}}{h_t} . \qquad (9)$$

These recursions are operational, since they involve only observable quantities.

Now all the ingredients of the optimization problem have been specified. As is obvious from (6), the period maximand has a multiplicative form in which the first two terms involve no parameters. The role of these terms is to introduce the Laffer restrictions given by (2). The third term corresponds to the evasion factor. It contains a parameter which is unknown and thus provides scope for learning. It also contains noise, thus learning is nontrivial. An alternative additive formulation with no noise in the evasion term but noisy observations on $R(r_t)$ is worked out in Appendix A.

First we establish that this is a well defined maximization problem. Note that given the distributional assumptions made, $\frac{\partial m_t}{\partial r_t} = 0$. This follows from (8) and (9) or, using a more general argument, from the martingale property of belief distributions generated by repeated application of Bayes' rule. The martingale property is derived for a more general formulation in Easley and Kiefer (1988) and will be encountered in the study several times.

Rewrite the period maximand as

$$E_t \ R(r_t) - \alpha \ r_t \ (1 - r_t) - m_t \ r_t^2 \ (1 - r_t) \qquad (10)$$

If (10) is strictly concave in r_t then so is (6). For this, the second derivative of (10) must be negative:

$-2\alpha - m_t[2-6r_t] < 0$ which holds if and only if[4]

$$\frac{\alpha}{m_t} > 3 \ r_t - 1. \qquad (11)$$

When passive learning takes place the policymaker maximizes (10) in each period, given current beliefs. Thus r_t satisfies the first order condition. Noting that one of the roots lies outside the admissible region for r_t given plausible values for α, the first order condition uniquely defines the optimal tax rate as:

$$r_t^* - \frac{\alpha + m_t - (\alpha^2 + m_t^2 - \alpha \ m_t)^{1/2}}{3 \ m_t}. \qquad (12)$$

We show that if (11) holds, then r_t^* in (12) achieves a unique maximum of (10), so that the problem is well defined in any single period.

Substituting (12) in (11), we have

$$\frac{\alpha}{m_t} > \frac{\alpha + m_t - (\alpha^2 + m_t^2 - \alpha m_t)^{1/2} - m_t}{m_t} \text{ , or}$$

$$\frac{(\alpha^2 + m_t^2 - \alpha m_t)^{1/2}}{m_t} > 0.$$

First the possibility of $m_t = 0$ is dealt with. If this were to occur, (11) is evidently satisfied and $r_t^* = \frac{1}{2}$ by L'Hopital's rule, therefore we still get a valid solution. If $m_t > 0$, on simplifying by m_t, we get

$$0 < (\alpha^2 + m_t^2 - \alpha m_t)^{1/2} = ((\alpha - m_t)^2 + \alpha m_t)^{1/2} \text{ .}$$

Since the square root is positive and by assumption $\alpha > 0$, the inequality will hold. This shows that if $m_t > 0$, the second order condition (11) will always be satisfied at the optimal tax rate given by (12). Hence $m_t > 0$ together with (11) is sufficient for the period maximand to have a unique maximum, i.e. the problem is well defined for an agent performing passive learning. If $m_t < 0$, the second order condition is never satisfied at $r = r^*$, since the inequality is reversed. Note that the quantity on the right hand side of the inequality is also the discriminant in (12) therefore r_t^* is always real for $m_t > 0$. Our problem is a well-defined one if $m_t > 0$ for the case of active learning as well.[5]

The conditions (11) and $m_t > 0$ need not always hold however, even if in the initial period they did. Sufficiently extreme realizations of the noise may result in a negative mean belief (this is most likely to occur in the first couple of periods for reasons highlighted in Appendix D) and this would mean that the problem is no longer well defined. Suitable choice of α

and β in the model can make the probability that this occurs very small but it cannot drive this probability to zero if the noise term has a distribution with infinite support. To rule it out, the double truncation of the support of u_t is necessary, as mentioned when u_t was introduced. This is not a restrictive assumption: for any of the parameter constellations utilized in the simulations (which will be described in Chapter IV), it implied a truncation affecting less than a percent of the probability mass of an untruncated normal with the same variance. It is interesting to note that the same end could have been achieved by different means as well: by applying a projection operator described in Appendix D. It should also be noted that a similar (though much milder) truncation actually occurs in any computer simulation study, since the absolute values of generated random variables can never be higher than the largest constant storable in memory, i.e. they also cannot take the values $-\infty$ or $+\infty$. Finally, though mathematically not posing a problem, the occurrence of $R(\tau_t) < 0$ ought to be ruled out because of the economic interpretation of this quantity. This is also achieved in Appendix D, and in effect it imposes an upper bound on the variance of the noise term.

Consider the form of the period maximand in (10). It consists of two parts: the first is deterministic, the second involves the unknown parameter at which learning is directed. The trade off between myopic optimization and experimentation is clearly present but now only a portion of the maximand can be affected by accumulating more information, the second term. This term involves the unknown parameter β. The problem is an extension of that considered in Prescott (1972). His results are drawn upon in this chapter.

We proceed by defining the value function as the function that gives the maximized value of the objective function in each period. It is obtained by plugging in the r_t's found optimal given the constraint $r_t \in [0,1]$ ∀ t and given current beliefs into the objective function:

$$V_t(p_t) = \max_{\substack{(r_s)^T_{s=t} \\ r_s \in [0,1]}} E_t \sum_{s=t}^T \delta^s E_s \left\{ r_s (1 - r_s) [\alpha - \beta r_s + u_s] \right\}$$

Clearly, the period t value function is a function of current beliefs $p_t \in P$. In general P can be thought of as the space of probability distributions with finite variance. For our model $p_t = N(m_t, h_t)$. Rewrite the value function as

$$V_t(p_t) = \max_{r_t \in [0,1]} E_t \left\{ R(r_t) + \delta V_{t+1}(p_{t+1}) \right\},$$

where p_{t+1} is obtained in the next period via Bayes rule involving p_t, r_t and e_t:

$$p_{t+1} = B(p_t, r_t, e_t).$$

Now all ingredients of the following form of the value function are defined:

$$V_t(p_t) = \max_{r_t \in [0,1]} \left[E_t R(r_t) + \delta E_t \left\{ V_{t+1}(p_{t+1}) \right\} \right]. \tag{13}$$

Given our formulation, the first term, current payoff, can be expressed in certainty equivalent form. The second term is the expectation of the attainable future maximum given current beliefs, Bayesian updating and future optimal behavior, including optimal experimental design. Thus the trade off between current gain and future information is present in this formulation. As is obvious from (8) and (9), the choice of r_t affects the posterior distribution, in particular, a higher value of r_t implies higher posterior precision.

II.2. Optimality of Active Learning in General

Proposition: Experimentation (i.e. active learning) is optimal if the value function is convex in beliefs and strictly convex for some periods.

Convexity basically yields the possibility of recouping currently foregone payoffs in the future in expected value terms. This follows from the definitions of convexity and of the value function as we now proceed to show. Convexity, via Jensen's inequality[6] implies the inequality below:

$$E_t V_{t+1}(P_{t+1}) \geq V_{t+1}(E_t P_{t+1}) = V_{t+1}(P_t). \tag{14}$$

The equality in (14) is an application of the martingale property of beliefs generated via Bayes rule. For active learning to be optimal, the inequality in (14) must hold strictly in at least one period, because then for that period we can write (14) as

$$E_t V_{t+1}(P_{t+1}) - V_{t+1}(P_t) > 0 . \tag{15}$$

This gap is the measure of the expected gains to be had from actively learning: when it is positive, expected reward given anticipated posterior beliefs exceeds certain reward given current beliefs (cf. equation (A8) in Appendix A). This completes the proof. Note that if the value function is affine in beliefs in each period, active learning cannot pay. This case arises for example when future beliefs are represented only by future mean beliefs and in this case the non-occurrence of active learning is a direct consequence of the martingale property.

The second term in (15) is the certainty equivalent value function. It is usually simple to obtain. This is however not the case with the first term. No closed form for this term is available in the general case. This restricts progress that can be made in deriving the quantities involved analytically. Numerical solutions may be obtained, however. Prescott (1972) for example uses a piecewise linear approximation to the value function (assuming a quadratic period optimand). Then starting from the terminal period and going backwards he employs grid-search to obtain the optimum value of the value function in each period. Since the second term in (15) is readily calculated analytically, a measure for the magnitude of the expected gains from active learning can be obtained.

It will be now argued that the assumption of Bayesian updating and a constant β together imply that the value function must be convex in beliefs. (Note: not necessarily strictly convex). These assumptions imply that any information on β, whenever acquired, will not be forgotten. Given the assumptions listed in section II.1, more information cannot reduce the attainable maximum, that is, more information can't hurt[7]. Given the above, the maximum nature of the value function delivers our claim, that is formally stated as a

<u>Lemma</u>: $V_t(p)$, $p \in P$ is convex.

It corresponds to Lemma B in Prescott (1972). The proof is relegated to Appendix C. This property as seen, plays a crucial role in making active learning optimal. In any specific instance, special care is of course needed to ensure that the value function is not affine in each period, since the Lemma does not preclude that. However, this would either correspond to the case when all learnable information is irrelevant or to the case where future beliefs are represented merely by their first moments in the value function. Thus an affine value function is not an interesting case. Apart from this possibility, a well formulated, sufficiently general problem from the class discussed in this study inherently has potential for optimal active learning as a consequence of this Lemma. The specific assumptions made about the families of probability distributions employed, the specification of the constraint, the value of the discount factor δ, the extent and variability of risk aversion, the length of the horizon in the problem interact to determine whether this potential can be realized or active learning is suboptimal (or ruled out altogether).

II.3 Optimality of Active Learning in the Model

To establish when active learning is possible in the model presented, a result in Prescott (1972) is utilized. It relies on the assumption of normality for beliefs and the properties of the resulting value function as a function of beliefs. It reduces the number of parameters characterizing beliefs to one. In what follows, only the approach will be shown, the algebra is relegated to Appendix D.

Rewriting (13) by inserting the expression for the maximand and without loss of generality explicitly making the value function depend on the parameters of the normal distribution characterizing beliefs instead of current beliefs themselves, we have:

$$V_t(m_t, h_t) = \max_{r_t} \left\{ \left[\alpha \, r_t \, (1 - r_t) - m_t \, r_t^2 \, (1 - r_t) \right] + \right.$$

$$\left. + \delta \left[V_{t+1}(m_{t+1}, h_{t+1}) \right] \right\} , \qquad (16)$$

where m_{t+1} and h_{t+1} are obtained using the recursive formulae following from (8) and (9) (given shortly, as equations (21) and (22)).

To proceed, note that

$$V_t\left(k \, m_t, \, \frac{h_t}{k^2} \right) = V_t(m_t, \, h_t) \qquad \text{for all } k \neq 0. \qquad (17)$$

The proof is relegated to Appendix C.

Setting $k = \pm \, (h_t)^{1/2}$ in (17) in each period we obtain a useful homogeneity property of the value function:

$$V_t(m_t h_t^{1/2}, \, 1) = V_t(m_t, \, h_t) = V_t(-m_t h_t^{1/2}, \, 1) , \qquad (18)$$

from which it is concluded that only the value of $s_t = m_t h_t^{1/2}$ plays a role in determining the value (18) can attain. Prescott (1972) refers to s_t as the location parameter and shows that it measures the degree of certainty of beliefs about the unknown parameter β. His interpretation is only valid if beliefs converge, which necessarily occurs in our model as a consequence of the martingale limit theorem - cf. Easley and Kiefer (1988). Thus, given our assumptions, the value function depends only on a specific, time-invariant function of the moments of the posterior probability distribution: s_t. Hence we need be concerned only with this value, which is a remarkable simplification. It has to be noted however, that even though the value function can be expressed as a function of the single parameter s_t, this parameter itself can only be updated using both the updated mean and precision.

Define

$$v_t(s_t) = V_t(s_t, 1) \tag{19}$$

and rewrite the Bellman equation (16) as

$$v_t(s_t) = \max_{r_t \in [0,1]} E_t \left\{ \left[\alpha \, r_t(1-r_t) - \beta \, r_t^2(1-r_t) \right] + \delta \left[v_{t+1}(s_{t+1}) \right] \right\}. \tag{20}$$

The rule for obtaining s_{t+1} needs to be specified. To obtain it, consider the present period, t and a period in the future: j. The following formulae follow from the update rules (8) and (9):

$$h_j = h_t + \sum_{i=t}^{j-1} r_i^2 \tag{21}$$

$$m_j h_j = m_t h_t + \sum_{i=t}^{j-1} r_i e_i . \tag{22}$$

Since $s_j = m_j h_j^{1/2} = \dfrac{m_j h_j}{h_j^{1/2}}$, we have

$$s_j = \frac{m_t h_t + \sum\limits_{i=t}^{j-1} r_i e_i}{\left[h_t + \sum\limits_{i=t}^{j-1} r_i^2 \right]^{1/2}} . \tag{23}$$

The properties of $E_t \left\{ \dfrac{\partial s_j}{\partial r_t} \right\}$ are derived in Appendix D. It is shown there to be an increasing function of r_t. Thus experimentation in the form of increasing the magnitude of the control variable delivers additional rewards in expected value terms in the next period. Appendix D also proves that this property carries over to the general $j > t$ case.

Finally some comments are offered on the variability of the sequence of controls under the assumption of different modes of learning. Given a fairly firm belief on the sign of the unknown parameter and the "bigger is better" property, an actively learning strategy will employ controls that are more variable than a passively learning one, to exploit this property. (Variability can be measured by the sample variance of the control variable.) This is a result amply referred to in the literature, usually shown to be true under the assumption of a normal probability distribution embodying beliefs on the unknown parameter (compare Grossman et al. (1977), for example). This result and its quite severe limitations are formally described in Appendix B.

The claim that the "bigger is better" property is not general is supported by the example in MacRae (1972), where allowing for active learning leads to controls which are actually less variable initially. The rationale behind this result is that a prudent (not excessively risk loving) economic agent will not increase the magnitude of the control variable applied when not even sure of the sign of its effect. The optimal strategy is to wait with experimentation until the sign of the effect of controls is determined reliably. This implies that first the agent will make sure that while the effect is unknown, it is kept at a level that avoids possible substantial damage. Then, once the agent is confident at least about the sign of the effect, more variable controls will be applied to efficiently trace out the magnitude of the unknown effect. (An analogous argument could be made when the agent is unsure of the functional form of the model underlying the environment he is facing.) An alternative cause for the breakdown of the bigger is better result can be the specification of the objective function: if it includes dynamics other than that via the evolution of beliefs under active learning, it may fail even if otherwise it would hold - cf. Section IV.3.1.

II.4 An Alternative Specification

This section employs a different specification of the multiplicative Laffer curve to illustrate the difference between reducible and irreducible randomness in a model of learning. It also provides an example when the cause for active learning not to occur is the choice of the probability distribution involved in the setup. This section represents a detour - we shall return to the original specification after this section.

In the general Laffer curve given by (1) and (2), let

$$Q(\tau) = \bar{Q} \ (\ 1 \ - \ \tau \ )$$

where $\bar{Q}$ is constant, without loss of generality set equal to 1 and

$$S(\tau) = \eta \ \tau + a$$

where the known constant a is large in a sense to be made precise shortly when η, the parameter through which randomness enters the problem, is described. In a one period problem the objective is to maximize tax revenue $R(\tau)$, thus the policymaker's period maximand is given by:

$$R(\tau) = \tau \ (1 \ - \ \tau) \ (\eta \ \tau + a)$$

Now we formulate the policymaker's problem as a multiperiod problem with finite, known horizon $T \geq 2$. The policymaker in period 1 seeks to maximize

$$\max_{\{\tau_t\}_{t=1}^T} E_1 \sum_{t=1}^{T} \delta^{t-1} \left\{ \tau_t \ (1 \ - \ \tau_t) \ (\eta \ \tau_t + a) \right\}.$$

Again the period objective function itself is static, the only connection between time periods is via the evolution of beliefs. The parameter η is assumed to be a random variable. This assumption introduces into the problem an element of irreducible randomness. In what follows let η denote the random variable in the evasion term and η_t its realization in period t. To make room for learning, the probability density function governing the random variable η is assumed to be not known exactly by the policymaker. In particular, the policymaker is assumed to be aware of the fact that η is uniformly distributed over the interval $[0,\omega]$ but does not know the precise value of ω. He has beliefs on the values ω can take however, embodied in $p_1(\omega)$, a Pareto prior distribution. The usual interpretation of the Pareto distribution is that it gives the probabilities of values taken by a random variable above a given threshold. Since here the policymaker knows that $\omega \geq$ max $\{\eta_1, \eta_2, \dots \eta_{t-1}\}$ at any time t, it is natural to use this distribution here to represent the policymaker's beliefs about ω. Thus the prior is given by

$$p_1(\omega) = \begin{cases} \dfrac{\alpha_1 \beta_1^{\alpha_1}}{\omega^{\alpha_1+1}} & \text{if } \omega \geq \beta_1, \\ \\ 0 & \text{otherwise,} \end{cases}$$

where the parameters characterizing the period 1, (i.e. initial) beliefs satisfy $\beta_1 > 0$, $\alpha_1 > 2$.

The first order condition to the single period problem is given by

$$3 \, \eta_t \, r_t^2 + 2 \, (a - \eta_t) \, r_t - a = 0$$

For any value of $a>0$ and $\eta_t>0$, this yields a unique solution for the optimal tax rate that satisfies the constraint $r_t^* \in [0,1]$:

$$r_t^* = \frac{\eta_t - a + (\eta_t^2 + a^2 + a\eta_t)}{3\eta_t}$$

The second order condition is satisfied if the second derivative is negative which occurs if

$$r_t^* > \frac{1}{3} - \frac{a}{3\eta_t}.$$

This is ensured if a is larger than η_t for all t, which is always true if the policymaker knows that $a > \omega$. Note that in principle knowing this could convey information on the value of ω. Since this will be ignored in what follows, it is assumed that a is so large that the error committed in doing so is insignificant.

The belief distribution represents incomplete knowledge that can be perfected over time, i.e. reducible randomness in the problem. It is important to note that learning is only capable of reducing this latter kind of randomness. Given our distributional assumptions, repeated application of Bayes' rule can be used to model learning by the policymaker: the Pareto distribution is conjugate. Of course to have a meaningful problem, we must make sure that η_t can in fact be treated as observable. This is established now.

The policymaker at any time t has a record of all previous r_t and $R(r_t)$ values assumed to be observed without measurement error. Then

$$\eta_t = \frac{1}{r_t} \left\{ \frac{R(r_t)}{r_t(1-r_t)} - a \right\}$$

which is uniquely obtainable from the observations the policymaker has. Thus η_t can be treated as observable in each period. Given this, the probability density function for ω (or equivalently: the parameters defining it) can be updated: $p_{t-1}(\omega)$ is updated to $p_t(\omega)$ after a time period has passed and a new sample element η_t has become available. The following update rules for the parameters can be used (DeGroot 1970, p172):

$$\alpha_t = \alpha_{t-1} + 1$$

$$\beta_t = \max \{ \beta_1, \eta_1, \eta_2, \cdots \eta_t \}.$$

Given the updated parameter values the predictive probability density function for η_{t+1} can be calculated. It is simply the weighted average of possible probability density functions of η_{t+1} where weights are assigned according to the current beliefs on ω, as embodied in the latest posterior:

$$g(\eta_{t+1} | \alpha_t, \beta_t) = \frac{\alpha_t}{(\alpha_t + 1) \beta_t} .$$

Passive learning is achieved by recalculating the whole sequence of controls for all remaining periods in each period based on current beliefs but applying only the first term in that sequence (sequential open loop control). The probability density function according to which expectation is to be taken in the objective function in each period when calculating the value of the control variables outstanding is given by the predictive probability density above. As time passes it is more and more likely that expectation in the objective function is taken with respect to the true probability density function of η. This corresponds to reducible randomness being diminished. Note that in this procedure certainty equivalence is applied since the objective function is linear in the sole unknown quantity and the period objective function is static. Note also that given the assumptions made the distribution of η and that of ω is independent of the values r_t, the control variable takes in any period. This fact rules out the possibility of active learning and leaves passive learning as the only option even if the policymaker seeks to optimally exploit all possibilities of gathering information available to it at every time point.

We now have two sequences of optimal controls: one corresponding to open loop control, $[r_1, r_2, \ldots r_T]^{OL}$ and the other one to sequential open loop control, $[r_1, r_2, \ldots r_T]^{SOL}$. The former solves (4) in each period with unchanging level of information on the distribution of ν and since the objective function is static, the solution is clearly a constant sequence. The latter is obtained from the sequential open loop procedure outlined above. Evidently it possesses higher variability, driven by the evolution of beliefs via passive learning.

This section has shown that by making a more structured set of assumptions on the unknown quantity, a distinction can be made between reducible and irreducible randomness in the problem. The distributional aspect of the setup is taken from DeGroot (1970, p172) and Crawford (1973). The specified period objective function is static, thus again the sole source of dynamics is that of beliefs. The distributional assumptions rule out active learning, a possibility well worth emphasizing. In the case of the previous specification, with beliefs represented by a normal distribution active learning is ruled out only if the restriction is applied that future beliefs are represented only by their first moments. Thus there it is the specification of the constraint that results in active learning being ruled out. Both cases are rather specific and in general, as argued in sections II.2 and II.3, active learning inherently has a role to play. Obviously, excessive discounting of the future is a third way to suppress this role.

II.5 Summary

This chapter introduced a model of Bayesian learning. The problem posed is well defined for all periods with passive and active learning if (11) and some additional conditions described in full detail in Appendix D are satisfied. Basically the conditions restrict the extent of uncertainty the decisionmaker may face. If uncertainty about the coefficient to be learned is too pervasive, then the sign of the expected return to experimentation (in our case, increasing the magnitude of the applied policy variable) can turn ambiguous. Thus (D6) is found to extend to a somewhat more general context the seemingly counterintuitive result in MacRae (1972) which has already been described. It thus represents evidence that MacRae's result is rather general.

The magnitude of the control variable may or may not be important in acquiring information through experimentation. In the cases when the bigger is better result (described in Appendix B) is true - as for the first but not the second specification of our model - active learning produces an initially more variable policy than passive learning or no learning at all.

The specification is not subject to the criticism that strategic aspects of the situation are unduly neglected as would be the case with the interpretation involving a monopolist experimenting to learn about the demand curve. This is because the government has an uncontestable monopoly as holder of tax levying rights, thus there is no other player who directly values the information on the unknown parameter of the model (a public good). Finally, it is argued that the other "player", the tax bearing public, is too atomistic and holds too disparate interests to form a stable coalition to engage in a conjectural variations game.

It is shown that in problems of the class considered here the possibility of active learning inherently exists but it is not necessarily optimal. Specific distributional assumptions or a restrictive specification for the objective function may rule it out altogether. Naturally, both active and passive learning are superior to open loop policy, i.e. totally myopic optimization.

The statistical procedure that the agent is endowed with in the model is more general than it seems in one respect, less general in another. Although Bayesian updating is utilized, it is well known that with a diffuse prior, a normal belief distribution to be updated and a squared error loss function, the result coincides with that of ordinary least squares learning. Therefore the model presented encompasses some cases of ordinary least squares learning as well. On the other hand, correct specification of the model is a maintained hypothesis not subjected to statistical tests. This leaves the setup open to the criticism that it is possible that a false model is accepted by the learning agent - an error analogous to type 2 error in statistical testing.

Notes

1. The double truncation of the support of the random variable u_t is meant in the sense that its support is $(-K, K)$, rather than $(-\infty, \infty)$ where K is a positive constant depending on the values of the parameters α, β, s^2 in the model. For reasonable parameter values K is always large. For a more formal argument consult Appendix D. For additional arguments in favor of this assumption, consult Kiefer (1988-89, section 4.)

2. This property will be repeatedly utilized: e.g. in (9) and in Appendix D.

3. The convention for subscripts is the following: m_t is the mean belief about β at the time when r_t is chosen, but before R_t is observed.

4. A regularity condition involving the conditional expectation of the unknown parameter is not usual. It could be replaced by appropriate technical conditions on the support and variance of the noise variable implying an m_t sequence satisfying (11) in each period.

5. In the sense that a maximum will exist, though it will no longer be necessarily unique. Active learning can pose further problems as well by generating noncausality in some formulations of the optimal control problem. Hughes-Hallett and Rees (1983, p.277), among others address the issue of the dynamic programming solution becoming suboptimal in the presence of noncausality or a non additively separable objective function. This does not invalidate our results however: even if the method of computing the maximizer (and hence the maximum) of the objective function is not dynamic programming, it can be computed (a method for doing so is presented in section IV). Thus the value function is still a well defined object and all the results to be obtained in this section using the value function remain valid. Moreover, Easley and Kiefer (1988) give a transformation of the value function following Bertsekas (1976, Chapter 4) which results in a formulation not subject to the Hughes-Hallett and Rees criticism.

6.   Jensen's inequality is used for a function of a probability distribution. The validity of this step follows from the fact that convex linear combinations of probability measures are also probability measures.

7.   The assumption that more information cannot hurt is not as innocuous as it seems. Although it definitely holds for our setting, for more sophisticated models counterexamples can be found. Arrow (1978) contains an especially simple and intuitive one relying on the fact that additional information may eliminate the possibility of trading risks without doing any offsetting good in a pure exchange economy.

III. Econometric Implications

Having formalized the concept of learning and the evolution of beliefs in a specific example we now proceed to show what the implications of learning for econometric practice are. Most of the concepts needed for this have now been defined, but some further technical econometric definitions will prove helpful by facilitating precise description of the effects.

III.1 Definitions

Concepts of exogeneity will be defined following Engle, Hendry and Richard (1983). In particular, to obtain an operative definition, Granger causality will be defined in a manner slightly different from Granger's original (population, rather than sample based) definition. This follows in spirit the operational definitions given in Granger (1980).

The ideas underlying the seminal treatment of the subject of exogeneity by Koopmans (1950) and the exogeneity definitions in Engle, Hendry and Richard (1983) are the same. The latter authors add some important refinements however. The underlying theme in these definitions is that the goal of econometric analysis with a given model must be clearly defined. Given this, a corresponding notion of exogeneity can be found, which, if valid for the particular variables in the model, facilitates achievement of this goal. Overall, the notion of exogeneity is geared towards allowing efficient analysis of relationships among a subset of variables without having to specify explicitly how the rest of the variables - those that are deemed exogenous - are generated. The results of the analysis are always conditional on the validity of the exogeneity assumption. The analysis of the variables takes a specific parametric form: a set of parameters of interest is chosen and exogeneity of variables is defined for this given set of parameters of interest.

Let us first define the usual concept of exogeneity and predeterminedness of a variable. For sufficient generality, consider a linear complete dynamic simultaneous equation econometric model with additive i.i.d. disturbances. The variable z_t in this model is (strictly) exogenous if it is uncorrelated with all current, past and future disturbances in the model. It is predetermined if it is uncorrelated with all current and future disturbances.

Engle, Hendry and Richard distinguish three distinct but interrelated goals in econometric analysis: inference, prediction and policy experiments. To each corresponds an appropriate exogeneity concept: weak exogeneity, strong exogeneity and super exogeneity, respectively.

Variables can have the property of weak, strong and super exogeneity depending on what the parameters of interest are chosen to be. Hence, as Geweke (1984) notes, these definitions depend on the loss function of the investigator and are in this sense subjective. From this alone it is also clear that none of these concepts is equivalent to strict exogeneity or predeterminedness.

Strict exogeneity and predeterminedness is neither necessary, nor sufficient in general for the goal towards which the Engle, Hendry and

Richard exogeneity concepts are geared: inference in models conditional on exogenous variables without loss of relevant sample information - cf. Engle, Hendry and Richard (1983)[1], Geweke (1984).

Now let us proceed to the formal definitions of weak, strong and super exogeneity. Given the econometric model, denote the parameters of interest by ψ and observed variables by $x_t' = [y_t'\ z_t']$. Parameters of interest are those parameters of the model which the investigator cares about. Note that no explicit restriction is made on what to include into the vector ψ. The joint density of the observations can always be factored as a product of a conditional and a marginal density:

$$D(x_t;\lambda) = D(y_t|z_t;\lambda_1)\ D(z_t;\lambda_2). \qquad (24)$$

If: (a) $(\lambda_1,\lambda_2) \in \Lambda_1 \times \Lambda_2$,

i.e. there are no cross restrictions between λ_1 and λ_2, or equivalently: this factorization "operates a cut",

and (b) $\psi = f(\lambda_1)$, i.e. parameters of interest can be uniquely determined from λ_1 alone,

then inference about ψ from the joint density $D(x_t;\lambda)$ is equivalent to inference about ψ from the conditional density $D(y_t|z_t;\lambda_1)$ alone. Therefore in this case no relevant sample information is lost by using the conditional density only: z_t can be treated as if it was determined outside the conditional model. This makes the analysis simpler - often vastly so.

Definition: If (a) and (b) hold, then z_t is weakly exogenous for estimating ψ.

Note that this definition does not preclude a relationship between lagged y's and z_t. If such a relationship exists, clearly one cannot take z_t's as **fixed**, only as determined outside the conditional model. This does not pose a problem for inference about ψ given a fixed sample. It does however, for prediction (which is always conditional on a set of fixed future values for the exogenous variables). Clearly, for valid prediction, a stricter definition of exogeneity is needed to also rule out the possibility of z_t's being affected by earlier y_t's. This additional requirement coincides with that of Granger noncausality from lagged endogenous to exogenous variables.

To obtain a formal definition, let X_0 denote the matrix of initial conditions taken as given,

$$X_t^1 = \begin{bmatrix} x_1' \\ \vdots \\ x_t' \end{bmatrix} \quad \text{and} \quad X_t = \begin{bmatrix} X_0 \\ X_t^1 \end{bmatrix} = \begin{bmatrix} Y_0 & Z_0 \\ Y_t^1 & Z_t^1 \end{bmatrix} = \begin{bmatrix} Y_t & Z_t \end{bmatrix}.$$

The process generating the sample of size T is represented by the joint density function $D(X_T^1 | X_0, \lambda)$, where λ is assumed to be identified.

Definition: Y_{t-1}^1 does not Granger cause z_t with respect to the information set consisting of X_{t-1} if and only if

$$D(z_t | X_{t-1}; \lambda_2) = D(z_t | Z_{t-1}, Y_0; \lambda_2). \tag{25}$$

If this holds for t=1,2,...T, then

(c) y does not Granger cause z.

Definition: If (a), (b) and (c) hold, then z_t is strongly exogenous for estimation of ψ.

Thus weak exogeneity of variables in the model and the lack of Granger causality from lagged endogenous to current values of these weakly exogenous variables together constitute strong exogeneity of these variables for the estimation of the parameters of interest. Strong exogeneity sustains prediction based on a __fixed__ set of forecasted future values for the strongly exogenous variables.

The third kind of use econometric models are put to, is conditional policy analysis. To sustain the validity of such exercises, the possibility of λ_1 (and hence of ψ) not being invariant to changes in λ_2 must be ruled out. In more general terms, the conditional density is to be invariant to changes in the marginal density, i.e. changes in regime. For us to have an operational notion, the class of regime changes considered must be clearly spelled out. For our purposes it is sufficient to formulate the requirement of invariance of the conditional density to changes in the marginal density as

(d) λ_1 is invariant to changes in λ_2.

Definition: If requirements (a), (b) and (d) hold, then z_t is super exogenous for ψ.

Note that Granger noncausality has been dropped from the list of requirements. If z_t is super exogenous for ψ, then once the parameters ψ in the model have been estimated, conditional policy experiments with λ_1 fixed yield valid results.

Two remarks are in order. First, the Lucas critique is the criticism of an unsupported assumption of super exogeneity of policy variables in macroeconometric models in the presence of forward looking expectations and the description of resulting simulation failures of the conditional model. We shall return to this phenomenon shortly.

Second, super exogeneity is always defined for a class of regime changes, and conditional policy experiments are valid only if the policy variables being shifted are super exogenous for a class of regime changes that includes these shifts.

III.2 Implications

Now we are equipped with all the tools, both economic and econometric, to make a number of points. In doing so, two situations must be clearly distinguished. The first one is the situation facing the policymaker which has been described in chapter II. The second is the situation of an outside econometrician who has access only to the data set containing the endogenous and exogenous (including policy) variables. We shall be concerned with this second situation in this chapter.

We seek to concentrate on the impact of learning performed by the policymaker. To do so, a situation is considered in which r_t would be exogenous in all three senses defined above given open loop policy. Then the impact of learning performed by the policymaker is analyzed on the components of each of the three definitions.

III.2.1 The Impact of Learning on Weak Exogeneity

Consider the problem of the econometrician observing the data generated by the environment discussed in chapter II. Suppose the list of variables in the model the econometrician has in mind coincides with the one there. Then the joint density of the data can be factored as:

$$D(R_t, r_t; \lambda) = D(R_t | r_t; \lambda_1) \ D(r_t; \lambda_2) \qquad (26)$$

where comparison with (24) reveals that y_t and z_t correspond to R_t and r_t, respectively in the model of chapter II. Let $\lambda_1' = [\alpha \ \beta \ \bar{Q} \ s^2]$. The parameter of interest is β. The other parameters in λ_1 were assumed to be known to the policymaker - the outside econometrician will be assumed to have no less information on this count. Clearly, more restrictive assumptions on the information available to the outside econometrician could lead to a violation of requirement (b) and hence render r_t not weakly exogenous. With this assumption however, requirement (b) is obviously satisfied since $\Psi = \iota' \lambda_1$, where the vector $\iota' = [0 \ 1 \ 0 \ 0]$.

Parameters of the process generating r_t are included in λ_2. Given our setup this process is driven by a maximum function depending on the current information set which in turn contains past R's and r's. The evolution of processed information at time t is summarized by (8) and (9). Clearly then, λ_2 includes at least some elements of λ_1. For example, beliefs (and hence, r's) depend on α and β, as is evident from (7), (7') and (9).

Thus the fact that the policymaker is learning implies an overlap (i.e. a cross - restriction) between λ_1 and λ_2, so requirement (a) is violated.[2] Therefore r_t does not remain weakly exogenous for estimating β for the outside econometrician if the data were generated by a learning policymaker. This in turn implies that it is also neither strongly, nor super exogenous for β. Weak exogeneity of r for estimation of β is lost because β can be more efficiently estimated if the generating process for r is included in a joint estimation procedure since this procedure also involves β and it also reveals how information is gathered by the learning agent. Disregarding this would result in loss of efficiency in estimating β.

To describe an alternative, indirect way in which learning can cause loss of weak exogeneity of policy variables, let us briefly consider a simplified variant of the model in Townsend (1983). Agents with disparate information are learning about their environment from data that is subject to specific and general shocks. Agent i observes only

$$z_{it} = z^*_{it} + v_{it} \qquad\qquad t=1, 2, \ldots, T.$$

Given his assumptions on the economy and modelling learning via Kalman filtering, Townsend concludes that v_{it} is serially correlated. Suppose that the individual's decision rule is

$$y_{it} = \beta\, z_{it} + \epsilon_{it} = \beta\, z^*_{it} + (\beta v_{it} + \epsilon_{it})$$

This implies under linear aggregation the relationship

$$y_t = \beta\, z^*_t + \tilde{\epsilon}_t$$

where $z_t^* = \sum_i z_{it}^*$ and $\tilde{\epsilon}_t = \sum_i (\beta v_{it} + \epsilon_{it})$.

Therefore the aggregate marginal model has error $\tilde{\epsilon}_t$ which is evidently also serially correlated due to the presence of learning. For simplicity assume serial correlation of order 1. We now utilize example 3.3 in Engle, Hendry and Richard (1983), which concerns a simple conditional model with serially correlated errors:

$$y_t = \beta z_t + u_t$$

$$u_t = \rho u_{t-1} + \epsilon_{1t}$$

$$z_t = \gamma y_{t-1} + \epsilon_{2t}$$

$$\begin{bmatrix} \epsilon_{1t} \\ \epsilon_{2t} \end{bmatrix} \sim \text{i.i.d. } N\left(\begin{bmatrix} 0 \\ 0 \end{bmatrix}, \begin{bmatrix} \sigma_{11} & \sigma_{12} \\ \sigma_{12} & \sigma_{22} \end{bmatrix} \right).$$

Let the parameters of interest be $\psi = (\beta, \rho)$. Consider the case $\text{cov}(z_t, u_t) = 0$. This corresponds to equation (52) in Engle, Hendry and Richard (1983) describing the complicated cross - restriction between (β, ρ) and γ which violates both requirement (a) and (b). The assumptions on the economy and the form of learning in Townsend (1983) and those on the econometric model in Engle, Hendry and Richard (1983) can be superimposed. In this case, the serial correlation in the errors of the conditional model generated by learning can be argued as above to imply loss of weak exogeneity under the above zero covariance assumption.

Evidently the autocorrelation coefficient appears in λ_1. A case can be made that it also ought to be incorporated into ψ, the parameters of interest. If ψ was not augmented with ρ - which is perfectly admissible under the Engle, Hendry and Richard definition of exogeneity - then ρ would not be estimated and the transformation utilizing ρ could not be performed. This in turn would mean that it would not be possible to estimate the non-augmented ψ efficiently using the conditional model. The weak exogeneity definition of Engle, Hendry and Richard is geared towards making efficient estimation from a given sample possible and at the same time it places no restrictions on the choice of ψ. As this example demonstrates however, ψ cannot be chosen completely arbitrarily if the original motivation is not to be abandoned. If a parameter is indispensable for efficient estimation of others which are included in ψ, then this parameter must also be included in ψ.

To summarize: weak exogeneity - and consequently strong and super exogeneity - of policy variables in an econometric model fails if they are chosen by a rational learning agent. Also, learning by itself may cause the errors in the conditional model to become autocorrelated, possibly causing loss of weak exogeneity via this channel. Finally, the choice of the parameters of interest cannot be completely unrestricted.

III.2.2 The Impact of Learning on Strong Exogeneity

We now return to the model of chapter II again. The failure of weak exogeneity already implies that strong exogeneity will fail. It is still worthwhile to discuss the impact of learning on the other ingredient of the definition of strong exogeneity: requirement (c). The dynamics induced by nontrivial learning over the sample period (or over a subset of it) is in the nature of Granger causality: R_t, the endogenous variable at time t affects the subsequent expectation operators by contributing a nonzero increment of information to the information set. Choice of r_{t+1} in turn is a result of the control rule involving the E_{t+1} operator which is conditional on the current information set. Formally we have

$$D(r_t | r_{(t-1)}, R_{(t-1)}; \lambda_2) \neq D(r_t | r_{(t-1)}; \lambda_2),$$

therefore it is clear that R_{t-1} affects r_t in the sense of Granger causality:

(c')   R Granger causes r.

Requirement (c) is directly contradicted by (c'), therefore we have yet another cause for r_t to be not strongly exogenous once open loop policy gives way to policy with learning.

Given the model of learning described in chapter II, the choice of r_t can be formulated as a function of previous errors in predicting $R(r_h)$, $h < t$, plus an error. Equation (6) in Sims (1977) describes a similar situation. There, Granger causality running from the endogenous to the exogenous variable is formally proven. The problem with using his proof here is that the process (R_t, r_t) would have to be assumed to be jointly covariance stationary, which it is not - as will be argued in section III.2.4. Also, he assumes linearity in the equation corresponding to the decision rule determining the choice r_t in our framework and this may contradict the optimality of r_t.[3] Hence the more general intuitive argument for the Granger causality from R to r used above.

The emergence of Granger causality running from R to r due to learning is a clear cut result. Since the concept of Granger causality stirred considerable controversy (cf. the exchange between Zellner, Schwert and Sims in Brunner, Meltzer (1979)), it is worth pointing out that it is used merely as a label for incremental predictive content and no claim is made on Granger causality representing or misrepresenting true causal links.

Newbold (1978) has shown (in a linear context) that empirical conclusions on the presence of Granger causality may be distorted when measurement errors are present in the data. Assuming that this carries over to the nonlinear case (e.g. regarding the linear as a local approximation to nonlinear functions), this cause serious problems in our case, since with nontrivial learning, noise is <u>necessarily</u> an element of the environment. Measurement error can be argued to be one specific source for this noise. Thus, if learning is relevant for a specific problem, empirical tests on Granger causality may face a pitfall following from the very nature of the data. However, as noted in Granger (1980), measurement error does not necessarily produce spurious Granger-causation. In fact, the only case when it does, is when the noise has a very particular time series structure.

III.2.3 The Impact of Learning on Super Exogeneity

In this section we focus on requirement (d) since as mentioned before, the failure of requirement (a) is already sufficient to render r_t not super exogenous. Practically, requirement (d) demands that for regime changes considered in conditional policy experiments, parameters of the conditional model can be treated as invariant to changes in the process generating the exogenous variables.

Invariance of structural parameters is an important issue. If a regime change occurs during the sample period and the parameters of interest are not invariant to it, then inference assuming constancy of parameters of interest throughout the sample period is invalidated.[4] If this regime change occurs during the forecast period with (d) not holding for it, prediction (utilizing the estimated parameters obtained in the earlier regime) is invalidated.

It will be argued now that learning in the data generating process implies loss of parameter invariance. The argument draws on the description of other mechanisms in the literature implying the same outcome, namely Lucas's (1976) and Geweke's (1985). Sargent (1981) makes the point that the observed behavior of economic agents changes if their perception of the constraints they face undergoes a change. The Lucas critique is aimed at instances when this occurs. Econometrically it amounts to saying that super exogeneity of policy variables may fail because agents - being in a game - adjust their expectations and hence their perception of the constraints they face. This results in a different optimal behavior for them. Since aggregated optimal decisions by agents constitute the data used in an econometric model, this fact renders parameters included in ψ dependent on regime changes for z. The effects of aggregation are ignored now, but it is important to note that implicitly assuming that the aggregator function is not sensitive to policy regime changes may be a mistake with consequences potentially as devastating as the modelling strategy criticized by Lucas: ignoring the potential sensitivity of expectations to changes in the policy regimes - cf. Geweke (1985).

This section seeks to demonstrate that a third, independent channel also exists which generates a shift in an agent's perception of the constraint he faces, hence implies a Lucas-type loss of parameter invariance. It is the effect of learning: the information of the agent on the constraint is augmented in each period (except for degenerate cases, such as the specification of section II.4). The key fact to note here is that the increment in information depends on the particular sequence of policy variables that have been applied. Thus, even in a non-game situation, and assuming away the potential sensitivity of the aggregator function to policy regime changes, learning behavior alone can explain a different observed behavior of the economic agent with a different data generating process for the policy variable. Hence, the assumption of learning is a plausible alternative source of loss of structural invariance of parameters in a model describing the behavior of an economic agent and thus has implications similar to those of the Lucas critique. A formal proof and comparison of the two related phenomena is now offered.

Sargent (1987, p.217) argues that with foresight, it will not be possible to find a representation expressing endogenous variables as a function of current and lagged exogenous variables: $y_t = f(z_t, z_{t-1}, \ldots, z_1)$, that is independent of the law of motion for the exogenous variables. Therefore, alterations in the law of motion for exogenous variables will alter the function $f(.)$. Hence parameters of this function cannot be assumed to be invariant to changes in the law of motion of exogenous variables.

Our case is analogous but slightly different in that it is the perception of the constraint of the agent whose behavior the $f(.)$ function describes that is in focus. Because of Bayesian learning, m_t, the policymaker's mean perception of the parameter β depends on all past values of r. Successive substitution into (8) and (9) yields (assuming for simplicity a diffuse prior):

$$m_t = \frac{\sum\limits_{i=1}^{t-1} r_i e_i}{\sum\limits_{i=1}^{t-1} r_i^2} .$$

Clearly therefore, (barring the unlikely occurrence of $\sum\limits_{i=1}^{t-1} \frac{\Delta m_t}{\Delta r_{t-i}} = 0$ for some t, where the Δ operator denotes the change in a variable) if the law of motion for r_t's - and hence their time path - was different, so would the corresponding m_t be. A different perception by the policymaker of the constraint he faces implies a different behavior for him (this will also be obvious from the formalization (28)). We then have the exact analogue of Sargent's formulation of the Lucas critique: the function describing the behavior of the policymaker cannot have parameters independent of the law of motion for exogenous variables.[5]

There are important differences between the effect of learning and the phenomenon in the focus of the Lucas critique. The first one is that learning as characterized in chapter II seems to have only limited impact since it yields beliefs that converge to a degenerate distribution with the whole probability mass concentrated at the true value of the unknown parameter. Therefore it would appear as if learning induced loss of structural invariance was a transitory phenomenon with no practical significance on the long run. There are several reasons why this is not the case. To begin with, what is exactly the long run in practice? If it means decades or longer, then obviously "transitory" phenomena are of interest. Also, beliefs converge to the truth with Bayesian updating only if the environment is stationary and uncertainty is limited. To the extent that the model describes reality closely by formally incorporating incomplete information of the agent about the environment, the Lucas criticism does apply, even though the highlighted phenomenon dies out asymptotically. Furthermore, there is evidence in the literature that even in a stationary environment beliefs need not always converge to the true value of the unknown parameter (McLennan (1987)). In fact the possibility of the limit of beliefs being different depending on the sequence of controls applied has been raised - cf. Kiefer (1988-89) and the example given in Feldman (1988-89). This phenomenon is peculiar to learning and it opens up the exciting possibility of a more fundamental breakdown of parameter invariance than the one discussed here. As opposed to a receding, asymptotically disappearing effect, a policy experiment applying a different set of controls risks that beliefs of the economic agent may converge to a limit completely different from the one they converged to with the actual sample sequence giving rise to the value of the parameters in the model. Then in general other (possibly vastly different) parameter values would be implied.

Kiefer (1988-89) also argues that convergence of parameter estimates by a learning agent need not occur in the econometric sense even if the environment is stationary. Finally, if the environment is nonstationary, learning will not necessarily recede and beliefs do not necessarily converge. In summary, the effects of learning cannot be treated as merely transitory phenomena that are asymptotically irrelevant.

Another difference between the effect of learning and the invariance phenomenon in the focus of the Lucas critique is that the data generating process for the policy variables is an element of the same optimization problem as the process for the endogenous variable in the former case. Hence there are restrictions on what other sequences of policy variables would qualify as a "learner-generated" sequence. Therefore, the class of admissible policy regime changes in this setting is restricted even before the question of super exogeneity of policy variables with respect to that class of regime changes can be posed. Clearly, some sequences of policy variables will not pass as "learner-generated". The restrictions placed by the assumption of learning on the path of the policy variables are relatively mild, however, as shown by the results of the simulation exercise reported in the next chapter and also by the examples on types of learning-induced regime changes given below. An example for this kind of restriction is the typical shape of control variable time paths generated by a Bayesian learning agent, discussed in the next chapter.

To see the role this restriction can play, consider a policy experiment with a proposed sequence of policy variables that has a monotonously increasing variance over the simulation period. If learning is hypothesized to play an important role during the sample period (observing a declining variance over time for the policy variable is an indication that it may), then even before the question of whether the policy variable is possibly super exogenous for the parameters of interest is posed, the policy experiment should be rejected because a learning agent can't have generated a control sequence with increasing variance. If the proposed policy variable time path is not incompatible with the assumption of learning, then the question can be posed, whether parameters of the model describing the behavior of the policymaker depending on the policy variable can be assumed to be invariant to this regime change if the policymaker has been learning. The answer to this question is no. The reason was described above: the proposed policy variable time path would have generated a different path of beliefs about the unknown parameter in the constraint of the policymaker, who would have chosen different optimal actions based on the resulting different perceptions.

As argued above, learning places restrictions on the policy variable paths. One of the reasons that these restrictions are not very strong is that learning is compatible with a rich variety of policy variable profiles: smooth, and possibly also abrupt changes in the time path of the policy variable can occur as a result of learning by a rational agent. Learning induces a gradual (smooth) regime change for the exogenous variable since the marginal distribution for the process generating the policy variable r_t is changing slowly. A switch of modes of learning would constitute an abrupt change of regimes for the exogenous variable: the MacRae result discussed in chapter II is an example for that. Another example can be constructed using the stochastic extension of the idea underlying J. Kiefer's golden section search.[6] It relies on the known unimodality of the Laffer curve and after two observations enables the policymaker to substantially truncate the support of the belief distribution on β. Truncation of the support of the belief distribution may imply a discontinuous change in the process generating the r_t's.

Again let us depart from our model for the rest of this section to show that learning can affect super exogeneity via an indirect channel as well. Simultaneous learning by several agents with disparate information can generate serial correlation in their forecast errors, as already argued in section III.2.1. There it was shown that this can imply serial correlation of the error term in the conditional econometric model describing the aggregate behavior of the agents. Formalize this as

$$
y \mid z \sim N(c + Bz, \Omega), \qquad \text{where } \Omega = \begin{bmatrix} 1 & \rho & \rho^2 \cdots & \rho^{T-1} \\ \rho & 1 & \rho \cdots & \rho^{T-2} \\ & & \ddots & \vdots \\ \rho^{T-1} & \cdots\cdots & & 1 \end{bmatrix},
$$

the covariance matrix arising from a serial correlation of order 1. Further suppose that this conditional model arose from a joint normal distribution of y and z, where both y and z are T-vectors:

$$
\begin{bmatrix} z \\ y \end{bmatrix} \sim N\left(\begin{bmatrix} \varsigma \\ \eta \end{bmatrix}, \begin{bmatrix} V_{11} & V_{12} \\ V_{21} & V_{22} \end{bmatrix} \right)
$$

This implies the following conditional and marginal distributions:

$$
y \mid z \sim N\left(\eta + V_{21}V_{11}^{-1}(z-\varsigma), \; V_{22} - V_{21}V_{11}^{-1}V_{12} \right)
$$

$$
z \sim N(\varsigma, V_{11}),
$$

in turn implying the regressions

$$y = [\eta - V_{21}V_{11}^{-1}\varsigma] + V_{21}V_{11}^{-1}z + \nu$$

where $\nu \sim N(0, V_{22} - V_{21}V_{11}^{-1}V_{12})$ and

$$z = \varsigma + \mu$$

where $\mu \sim N(0, V_{11})$, respectively.

Equating quantities in the two alternative formulations of the conditional model we obtain the implied restrictions:

$$c = \eta - V_{21}V_{11}^{-1}\varsigma \;;$$

$$B = V_{21}V_{11}^{-1}$$

(since the conditional model should hold for any z);

$$\Omega = \begin{bmatrix} 1 & \rho & \rho^2 \cdots & \rho^{T-1} \\ \rho & 1 & \rho & \cdots & \rho^{T-2} \\ & & \ddots & & \vdots \\ \rho^{T-1} & \cdots & & 1 \end{bmatrix} = V_{22} - V_{21}V_{11}^{-1}V_{12} \;. \tag{27}$$

Let us now examine the Engle, Hendry and Richard definition of super exogeneity in this concrete setting. Define the vectors λ_1 and λ_2 and the parameters of interest ψ as:

$$\lambda_1 = (\, c,\, B,\, \Omega\,)$$

$$\lambda_2 = (\, \varsigma,\, V_{11}\,)$$

$$\psi = (\, c,\, B,\, \rho\,).$$

It has been argued before that given the other parameters included in ψ, inclusion of ρ in the parameters of interest is necessary to avoid contradicting the original motivation for the exogeneity concepts. Clearly, ψ can be obtained from λ_1 by exclusion of elements, therefore requirement (b) is satisfied. Let us first consider the restriction connecting the coefficient matrices in the two representations of the conditional model: $B = V_{21}V_{11}^{-1}$. From this it follows that in general, super exogeneity of z for ψ cannot be guaranteed since a change in V_{11} could well mean a change in λ_1 and hence in ψ as well. We are after a stronger result however: that in the setup given, super exogeneity necessarily fails. To see this, focus on (27), the restriction connecting the covariance matrices in the two representations of the conditional model. Also, assume $y_t \sim$ i.i.d. $N(\bar{y}, \sigma_y^2)$ and $z_t \sim$ i.i.d. $N(\bar{z}, \sigma_z^2)$. This rules out the possibility of serial correlation being "inherited" from either y or z. Then (27) becomes

$$
\begin{bmatrix}
1 & \rho & \rho^2 \ldots & \rho^{T-1} \\
\rho & 1 & \rho \ldots & \rho^{T-2} \\
 & & \ddots & \vdots \\
\rho^{T-1} & \ldots\ldots & & 1
\end{bmatrix}
= \sigma_y^2 I - \sigma_z^{-2} V_{12}'V_{12} \; .
$$

From this form it is obvious that if σ_z^2 (an element of λ_2) changes, then so does ρ (an element of λ_1 and also of ψ). Hence z cannot be super exogenous for the ψ specified. However, this is not necessarily a cross - restriction, therefore weak exogeneity need not fail.

The argument given for violation of requirements (a) and (d) are analogous. In fact they stem from the same root. First, let us explore the relationship between these two requirements. Clearly if an overlap between λ_1 and λ_2 occurs, then a change in one necessarily implies a change in the other. The converse, that if no overlap occurs, no change in λ_1 is implied if λ_2 changes, is not true, therefore requirement (d) is necessary. What it asks for is not merely that no parameter should appear in both λ_1 and λ_2, but more: that there be no (stochastic or deterministic) functional relationship between the elements of the λ_i's. In our case however, we have that λ_1 and λ_2 have common elements because of the feedback introduced by learning. This in itself is sufficient to violate both requirements (a) and (d).

The result that the presence of informational feedback implies an overlap between the two sets of parameters is generalizable to other kinds of feedbacks. An example would be a central bank committed to an interest rate peg in a stationary, stochastic environment. Money supply then clearly cannot be treated as weakly, strongly or super exogenous in a model including the interest rate as an endogenous variable.

It will prove useful to also explore the consequences of learning behavior in the data generating mechanism on the usual notion of strict exogeneity and predeterminedness. The argument on r Granger causing R is only sufficient for strict exogeneity to fail if the informational feedback was instantaneous. However, in keeping with Granger's and Engle, Hendry and Richard's definitions, and for reasons of tractability, the information feedback in this paper was assumed to occur with a lag. This leaves r_t being contemporaneously uncorrelated with an i.i.d. disturbance term, hence strict exogeneity need not be lost because of learning. This is not a contradiction with the earlier results: it merely represents the fact that consistent estimation of the parameter of interest may still be possible even though its efficient estimation is not.

The belief distribution need not always converge to a point mass. In case it does not, the policy variable can be nonstationary. This would invalidate an approach relying on the notion of consistency. In our simple model, nonstationarity of the policy variable tapers off as the belief distribution converges to a point mass. Even under these circumstances however, the decision period must be no shorter than the observation period for lagged information feedback to not appear as an instantaneous one in the data, which would imply a correlation between the disturbance term and the policy variable, i.e. a loss of strict exogeneity. Moreover, both r_t and the disturbance term u_t are functions of r_{t-1}, therefore they cannot be assumed to be uncorrelated in general. Again, in the simple model employed in this paper, this feature is only temporary, as the belief distribution collapses, the correlation induced by learning vanishes. Thus in the limit, the correlation is 0. This shows that strict exogeneity is a very different notion from the Engle, Hendry and Richard exogeneity concepts. Since asymptotic properties are not in the focus of this paper, it is only pointed out that the phenomenon of incomplete learning could very well lead to failure of strict exogeneity.

Now: why does the outside econometrician not extend his model to include the informational feedback? Joint estimation of the parameters of a model characterizing learning and those of the econometric model proper could be efficient. The basic obstacle to this is lack of observability of crucial variables necessary to formulate an identifiable model for learning. Beliefs are unobservable, and so are the following: degree of risk aversion, the precise information structure, the method of learning, the occurence and extent of forgetting, computational constraints, precise degree of rationality, the utility attached to acquiring information, etc.

This incomplete list is enough to reveal the complicated nature of the problem. Any aspect can be in principle quantified for inclusion in the model, but for this to be feasible, all the other aspects must be suppressed, i.e. assumed unchanging. Ceteris paribus is not a good way to approach this problem however, because there are strong interconnections. The appropriate notion of rationality for example depends on what the computational and data storing constraints are, and what variables are in the set of relevant information. Limits to observability affect this information set, which in turn conditions the method of learning applied (e.g. qualitative information cannot be incorporated the same way as quantitative information). Risk aversion affects the mode of learning. Strategic interactions may result in non Pareto optima equilibria, implying patently non-rational outcomes when viewed from a purely decision-making viewpoint.

In any case, a complex nonlinear system with numerous nonlinear, cross equation restrictions would need to be appended to the outside econometrician's original model. Quite aside from possible problems of identifiability, estimation would be disastrously nonrobust to changes in any of the assumptions made, due to the interrelationships detailed above. Thus for practical purposes, achieving efficiency of estimation by incorporating the learning mechanism in the outside econometrician's model is not feasible. This fact gives weight to the findings on loss of exogeneity above.

As a concluding remark it is mentioned that it is theoretically possible to conduct sound policy experiments even in the presence of structurally non-invariant parameters. Suppose the function describing the change in λ_1 induced by the change in λ_2 is known (or estimated on the basis of a hypothesized structure for the mechanism giving rise to this phenomenon): $\lambda_1 = \Gamma(\lambda_2)$. Then this change can be accounted for when interpreting the results of the policy experiment and meaningful quantitative results can be derived through a suitable standardization. Sargent's formula for the cross-equation restrictions implied by forward looking expectations on parameters of the causal representation of a rational expectations model can be interpreted as an example of the function $\Gamma(.)$ referred to above - cf. Sargent, (1987, p.216). With passive learning $\Gamma(.)$ could be constructed once a structural model is specified utilizing the decision rule (12) where mean beliefs are substituted out using equations (8) and (9).

If it were firmly believed that the formalization utilized to obtain the function $\Gamma(.)$ is valid, then the proposed standardization could be performed. This would salvage conditional policy experiments even in the presence of structurally non-invariant parameters (with learning, requirement (a) is still violated, though). In effect this way of handling the problem amounts to the following strategy. Because of the loss of invariance for the structural parameters the conditional model cannot be used by itself. Instead of utilizing the joint likelihood function of the observations however (which may prove to be computationally infeasible) a relationship is derived from additional hypothesized structure on the joint distribution that formalizes the impact of changes in λ_1 on λ_2. Substituting in $\lambda_1 = \Gamma(\lambda_2)$ instead of $\lambda_1 =$ constant into the conditional submodel alleviates the problem as long as the hypothesized structure yielding the

$\Gamma(.)$ function itself is safely assumed to be time invariant. It is obvious that for this approach to make sense, $\Gamma(\lambda_2)$ should yield λ_1 - constant for the policy regime during the sample period and λ_1 - another constant for the alternative policy regime. Although this approach only amounts to pushing the question of structural invariance back one stage, it may be able to cope with a number of specific cases.

Whether or not this approach is used, learning is a distinct source of structural non-invariance, and all the points raised about the Lucas critique equally plausibly arise if a learning agent is present in the data generating process. The only case when they don't is when beliefs don't matter: the decision rule is practically insensitive to changes in beliefs. As the simulations show, almost such is the case for our specification of the model with passive learning only, for some settings of the parameters α, β and s^2 (e.g. α - 8, β - 2, s^2 - 2).

III.2.4 Learning Induces Nonstationarity

Learning, whether passive or active, results in a nonstationary process for generating policy variables. To see this, it is enough to note that in a well defined problem Bayesian learning implies a strictly nested sequence of information sets $\{I_t\}$. Assuming that preferences of the agent do not change (and thereby avoiding a possible identification problem for the outside econometrician), expanding information sets imply that the function

$$\text{argmax} \left\{ E_t \ R(\tau_t) | \ I_t \right\} \tag{28}$$

will change as time proceeds (where R(.) denotes the maximand of the agent). The reason for the change is that (28) is a function of the information set, the change in which results in a change of (28), unless all the increment in information is irrelevant to the maximization problem. Since the agent always chooses τ_t to be the maximizer of the objective function given current information, the implication for the observed time series $\{\tau_t\}$ is that its moments are generically nonconstant - cf. the arguments presented on the variability of τ_t's at the end of chapter II and the description of implied regime changes in the previous section. Since learning is not always complete, i.e. it need not always result in point-mass final beliefs, this can be true in the limit as well. Note that in our concrete problem, since a static period maximand is assumed, the issue of stationarity boils down to whether or not $\{\tau_t\}$ is a constant sequence (as it would be under open loop policy).

III.3 Summary

This chapter explored the econometric implications of learning behavior by a policymaker who maximizes the discounted sum of tax revenues subject to the constraint imposed by a Laffer curve involving a parameter about which Bayesian learning occurs. In particular, the issue of exogeneity of policy variables has been addressed. The exogeneity definitions of Engle, Hendry and Richard of weak, strong and super exogeneity all fail to fit the policy variable as soon as learning about the constraint faced by the policymaker occurs. A restriction on the choice of parameters of interest in the definitions of exogeneity by Engle, Hendry and Richard was proposed, namely that parameters in the conditional model that are necessary for efficient estimation of any parameter in ψ must not be excluded from ψ. The analogy between the effect of learning and that of forward looking rational expectations on the structural invariance of parameters - the essential feature underlying the Lucas critique - was described. Learning has been identified as a possible source of nonstationarity of economic time series. Finally, the impact of learning on strict exogeneity has also been summarized.

Notes

1.   The reason they cite for this is that assumptions about the parameters of interest are not made explicit and thus the requirement that parameters of interest are functions of only the parameters of the conditional model is missing from the usual definitions. If this is not made sure then loss of relevant sample information cannot be ruled out when only the conditional model is used. In other words it may be the case that the marginal model contains non - nuisance parameters as well. Engle, Hendry and Richard (1983) contains several examples to highlight this point.

2.   More generally, any kind of feedback from observations of the endogenous variable to the function determining consecutive exogenous variables implies an overlap.

3.   C. Gourieroux, personal communication.

4.   It is a different issue whether this can be ascertained solely from the data. It may therefore have to be a maintained hypothesis.

5.   A similar point in a somewhat different context was made in DeCanio (1979, footnote 2).

6.   Suppose the policymaker knows for sure that the Laffer curve is stable and has a unique peak. Then, after 2 observations on r_t and R_t, he can rule out a large interval for the parameter determining the location of the peak. To see this, take $r_1 < r_2$ (relabel, if necessary) and denote the argmax by r^*. Then,

$$\text{if } R_2 < R_1 \text{ then } r^* \leq r_2 ,$$
$$\text{if } R_2 = R_1 \text{ then } r^* \in [r_1, r_2] \text{ and}$$
$$\text{if } R_2 > R_1 \text{ then } r^* \geq r_1,$$

with the implications for the support of the unknown parameter usually easily computable. For a discussion of the stochastic extension see LeCam and Olshen (1985).

IV. Simulation

This chapter describes a simulation exercise based on variants of the first specification of the theoretical model presented in chapter II. Following the description of the design and the goals of the simulation, the algorithms for obtaining the passive learning and active learning sequence of controls are described. The insights yielded by the exercise are then presented separately for the case of passive and active learning. The software used was Gauss 1.49B.

IV.1 Design

The model for the generation of revenue is:

$$R(r_t) = \bar{Q} \, r_t \, (1-r_t) \, [\alpha - \beta \, r_t + u_t] \qquad\qquad t=1,2,\ldots,T \qquad (29)$$

where $u_t \sim$ i.i.d. $N(0,s^2)$,

$\bar{Q}$, α, s^2 and T are known, positive constants;

r_t is the policy variable, and

β is an unknown constant.

Beliefs on the unknown parameter β are assumed to be embodied in a normal probability density function. For simplicity and without loss of generality in what follows, it will again be assumed that $\bar{Q} = 1$.

The overall constraint that beliefs in each period must satisfy for the problem to remain well posed, spelled out in detail in Appendix D and labelled (D6), was imposed in the actual simulations by choosing α to be large relative to the value of the parameters β and s^2.

Fixing a specific prior probability density function, the evolution of beliefs depends on the mode of learning (active or passive), and realizations of the random noise component u_t. Simulation in this context means generating many realization paths for u_t: $\{u_t\}_{t=1}^{T}$, each of which corresponds to a possible "world" in which the policymaker is learning about the Laffer curve. Fixing the prior probability density function and the mode of learning, in each of the possible "worlds" we get a path of beliefs, the implied path of controls and the discounted sum of payoffs.

The following aspects of the problem are of interest:

(A) How does learning proceed in a typical realization?

(B) How sensitive are results to different components of the problem?

(C) Does active learning yield paths of tax rates and beliefs different from the passive learning case?

(D) How is econometric estimation affected?

As we proceed to the special cases of passive and active learning, answers will be sought to these questions. Discussion of (D) is dispersed throughout this and the next chapter.

IV.2 Passive Learning

As before, denote current beliefs at time t by p_t. With passive learning only, the policymaker solves a static maximization problem in each period, only the information set available to him changes from period to period. Given that the problem does not cease to be well defined in any period, the policy variable chosen in period t will be

$$\tau_t = \text{argmax} \left\{ E_t \, R(\tau_t) \mid p_t \right\}. \tag{30}$$

Using the current certainty equivalent for β is valid since the maximand is linear in the unknown quantity: β. Then the optimal value of the policy variable is given by (12). Since the update rules (8) and (9), or equivalently, (21) and (22) completely define the transition rule for beliefs, once u_t and τ_t are given for each period, the evolution of beliefs is simple to compute. Thus for the case of passive learning the design of the simulation exercise is simple and can be summarized as follows.

Inputs: Prior probability density function and $\{u_t\}_{t=1}^{T}$, the hypothetical realization path of the noise variable

Outputs: the resulting sequence of optimal tax rates: $\{\tau_t\}_{t=1}^{T}$,

the implied sequence of revenues: $\{R(\tau_t)\}_{t=1}^{T}$,

the implied sequence of beliefs: $\{p_t(\beta)\}_{t=1}^{T}$ and

the resulting total discounted payoff: $\sum_{t=1}^{T} \delta_t R(\tau_t)$, where δ_t is the discount factor. For all results presented in this chapter, the values for the parameters were: $\alpha = 1000$, $\beta = 900$, $s^2 = 15000$. For reasons that will become transparent in section IV.3, a dynamic version of the maximand containing a lagged revenue term has also been employed in the simulations. In this section, the dynamic objective function results are computed for comparison purposes only. The precise nature of the intrinsic dynamics together with motivation and the presentation of the numerical optimization algorithm utilized in the computations will be given in section IV.3.

IV.2.1 Results - Passive Learning

Even though this is a very simple exercise, it provided essential (though sometimes perhaps obvious) insights into the nature of the problem, summarized below.

(a) Technical details of the model:

The constraint (D6) can be quite restrictive for some parameter constellations even for relatively small variance of the noise term. To ensure that it holds either both α and β had to be large, or if both were small, the value of α had to be much larger than that of β. In either case the variance of the noise term could not exceed the bound imposed by (D6). It turned out however, that the cases for which problems with (D6) were likely to occur were exactly the ones that proved to be uninteresting in the sense that beliefs had practically no impact on the choice of optimal tax rates. This phenomenon was caused by the $r_t(1-r_t)$ term in the maximand: if β was low, in any run that did not violate (D6), the optimal tax rate always stayed in the close neighborhood of $\frac{1}{2}$, the value maximizing this term. Note that (D6) restricts the magnitude of s^2, hence the possible bias in any period after the first few. More interesting results were obtained when α and β were large (α had to be large to comply with (D6) once a large β was chosen). This allowed learning to become more important but even so, uncertainty (including both uncertainty about the value of β and the magnitude of the variance of the noise component) could not be allowed to be too high. If it had been too high, it would have implied a structural break: a cautious policymaker would over time be replaced by a more experimenting one for reasons discussed in MacRae (1972): having become sure of the sign of the effect, the policymaker begins vigorous learning to trace out its magnitude. The model used is not rich enough to capture this kind of behavior: the problem may become ill defined for the case which would produce the first kind of behavior of the policymaker. It is worthwhile to note however that the above described phenomenon is in principle capable of generating a structural shift. In empirical work therefore, if learning is suspected to have played a significant role in the data generating process, application of Chow-type diagnostic tests is advisable - cf. Chow (1960).

Given the assumption of normality, the characterization of how beliefs evolve need only involve the mean and the precision of the belief distribution. Illustrative paths of mean beliefs are presented in alternative situations in graphs 1 and 3 for the static objective function case and in graph 2 for the dynamic objective function case. Mean beliefs, as it can be seen on the graphs, followed a typical path: after an initial couple of periods when hectic jumps and sometimes reversals of the bias relative to the true value of the unknown parameter occurred, they converged rather smoothly to the true value. It is interesting to note that the initial hectic jumps occurred even if the policymaker was endowed with the true value of the parameter β as the initial prior mean. This effect was of course due to the fact that initial precision in these runs was always near zero. The magnitude of the jumps depended on the variance of the noise variable. The bias did not change signs if $|\beta-m_1|$ was excessively large - in this case convergence usually occurred from one side, quite smoothly.

In the case when initial precision was set at a high level, changes in mean beliefs became remarkably smooth. This is an advantage only if initial beliefs were correct since otherwise smoothness in mean beliefs amounts to stubbornly sticking to beliefs that are repeatedly proven incorrect period after period. If initial mean belief was set to be substantially lower than the true β, $h_T - h_1$ was an increasing function of h_1, and when m_1 was much higher than β, it was a decreasing function of h_1. Given (21), this clearly implies that in the former case tax rates tended to be chosen to be larger than $\tau_t^{opt}|_{m_t=\beta}$ and conversely in the latter case.

In this simple model, the sequence of mean beliefs always converges to the true β value, regardless of the (positive) value chosen for m_1. The speed of convergence depends on the variance of the noise term and on how correct and tight prior beliefs are. Graph 4 shows the effect of the tightness of the prior for the static objective function case when initial mean belief was set to half the true value of β: the numbers in the variable names refer to the value of the initial precision. The number of time periods after which convergence of mean beliefs occurs depends primarily on the variance of the noise term and on initial precision. For some constellations of the parameters, it was found to be as low as 5 to 10, using other constellations however, it could be made to be well over 100.

Allowing for intrinsic dynamics in the objective function has some interesting effects on the passive learning algorithm, but it does not change the overall picture. Convergence of mean beliefs to β becomes somewhat slower and results somewhat more sensitive to the variance of the noise term. The reason is that because of the presence of the lagged revenue term, the period objective function becomes a function of all past disturbances as opposed to the case of the static period objective function, which depends only on the current noise term. The numerical optimization algorithm converged to the same optimal solutions as long as initial values for the tax rates were set to be higher than about 0.3. Otherwise it invariably failed to converge in at least one of the periods or yielded wildly differing solutions that were often very unreasonable. The conclusion drawn from this is that above this threshold the algorithm converges to a locally unique and economically interpretable optimum. In the active learning algorithm specifying initial tax rates to be above this threshold proved to be a safe bet as well.

There is no qualitative difference between diffuse and nondiffuse priors except for one aspect: analytical derivations utilized in the part describing the algorithm to obtain active learning controls will not go through if the prior is diffuse, i.e. $h_1 = 0$. This is not a great problem since after an initial period, a completely diffuse prior is updated to one with nonzero precision. Then treating the problem as starting off from that period solves this problem. If the initial mean belief is very much off the mark, a sufficiently high initial precision will make beliefs differ from the truth substantially even after 50 or 100 periods, as already pointed out. This cautions against using too tight priors in the initial periods when modelling learning. This is not a severe restriction since the only case when a very tight prior can be assumed is when it can be hypothesized that the agent is not far from actually knowing the true value of the parameter he learns about. This however is exactly the case when learning is unimportant.

Beliefs are in general unobservable, but tax rates are observable. Therefore, from the econometrician's point of view it is of more interest how the tax rates corresponding to the reported beliefs evolved. This is discussed next.

(b) Econometric aspects

Optimal tax rates corresponding to mean beliefs reported in graph 4 are displayed in graph 5. Clearly, despite the fact that mean beliefs are initially quite unstable, the path of optimal tax rates is found to be relatively smooth. Graphs 6 and 7 confirm this conclusion: they display two more paths of tax rates (initial precision is 0 in both cases, initial mean belief is $\beta/2$ for graph 7). The smoothness of the path of tax rates is not a generalizable property however, since it is a consequence of the specification: the maximand involves the dampening $\tau_t(1-\tau_t)$ term. On the other hand, this specification is not completely arbitrary: it is necessary to comply with the requirement that the problem be well defined in all periods as beliefs evolve. Including this term may actually be advantageous, because as described in the next section, it may have contributed to avoiding problems of nonuniqueness of the optimal solution path.

All in all the effects of passive learning on the observable variables were found to be less than overwhelming after the first couple of periods in the model of chapter II with the parameter values tried. With a nondiffuse prior however, if the initial mean belief was incorrect, a stubborn positively serially correlated bias was introduced. The first couple of controls in the case of a diffuse prior were found to be markedly different from the rest. This phenomenon will also be seen to occur with active learning.

Different possibilities would open if $(u_t)_{t-1}^T$ was not assumed to be identically and independently distributed. For example, it could be autocorrelated, or it could have a mean or variance shifting over time. If the random noise component was serially correlated then its future realizations could be forecasted and the controls for the future could be chosen taking this into account - an instance of feedforward control. The simple interpretation of the expectation operators in (6) would no longer hold, however. In any case, this subject is not pursued further here.

Graph 1

Convergence of Mean Beliefs with Diffuse Prior and
Different Initial Means under Passive Learning

Static Objective Function Case

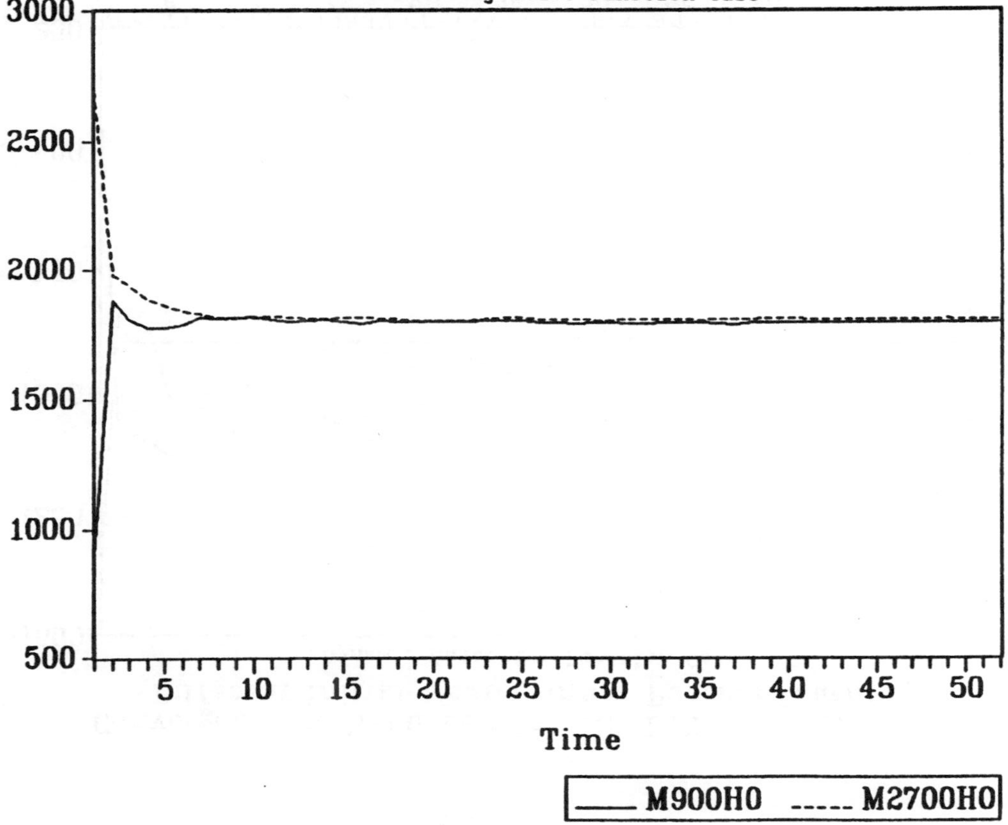

Graph 2

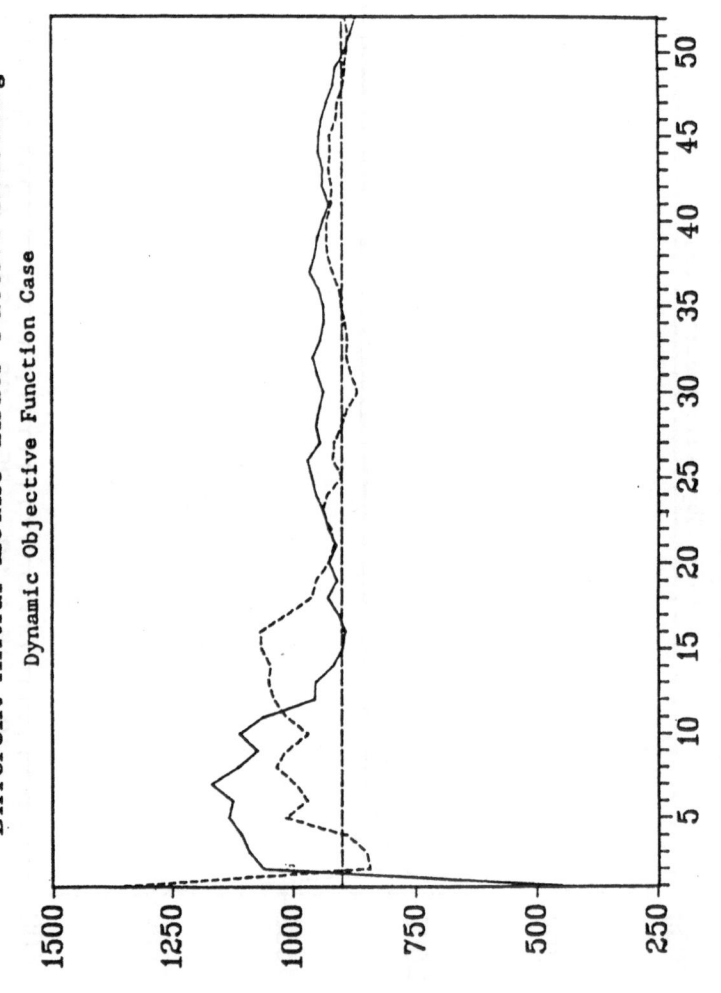

Convergence of Mean Beliefs with Diffuse Prior and Different Initial Means under Passive Learning

Dynamic Objective Function Case

M450HOD ———   M1350HOD ‑‑‑‑‑   M900 ‑‑‑‑‑

Graph 3

Mean Beliefs with Diffuse Prior when Initial Mean Equals True Beta under Passive Learning

Static Objective Function Case

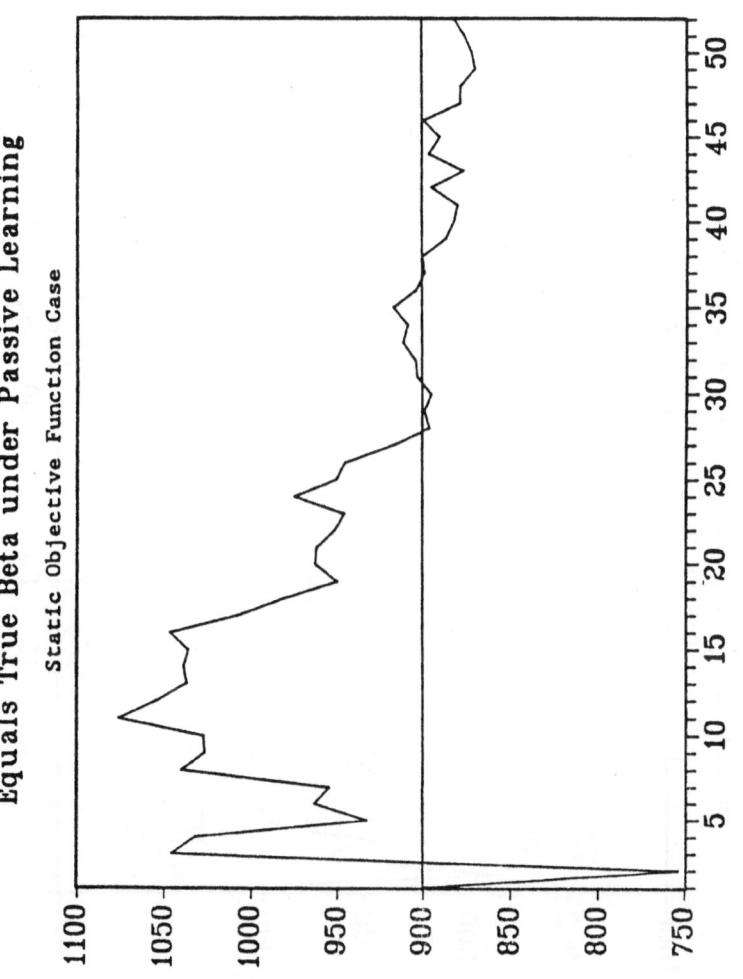

MHOBETA

Graph 4

Convergence of Mean Beliefs with Different Values
of Initial Precision under Passive Learning

Static Objective Function Case

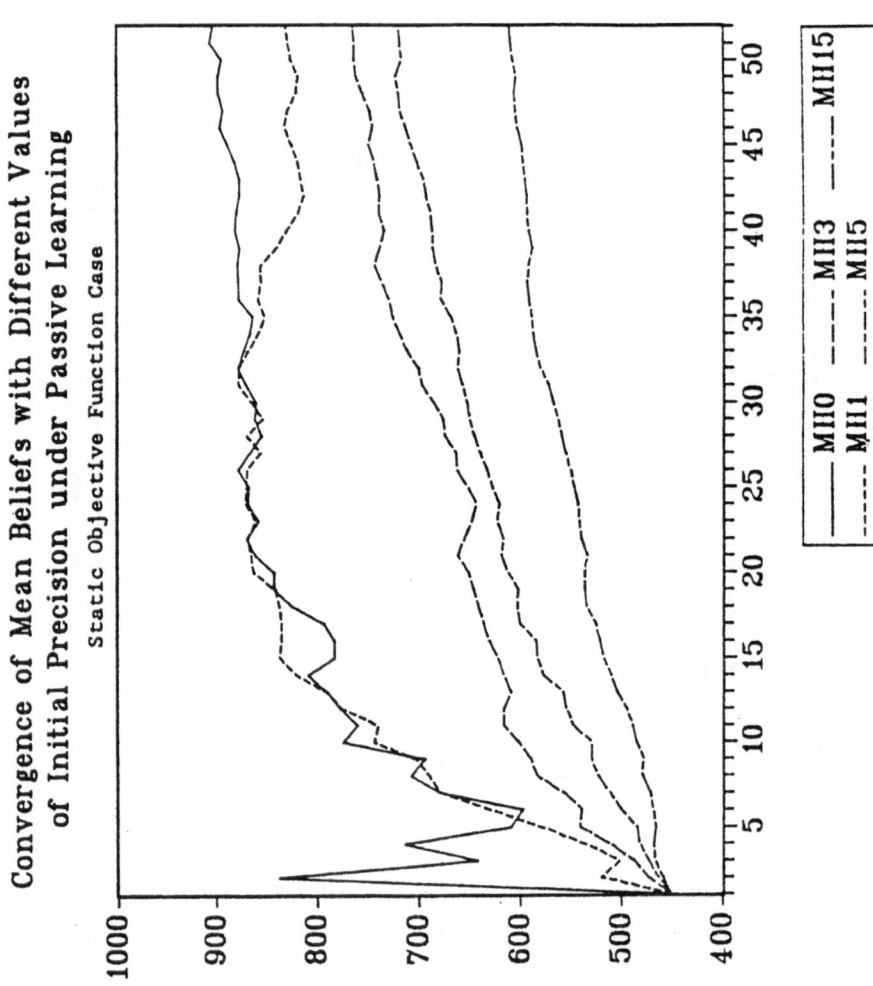

Graph 5

Sequence of Optimal Tax Rates with Different Values of Initial Precision under Passive Learning

Static Objective Function Case

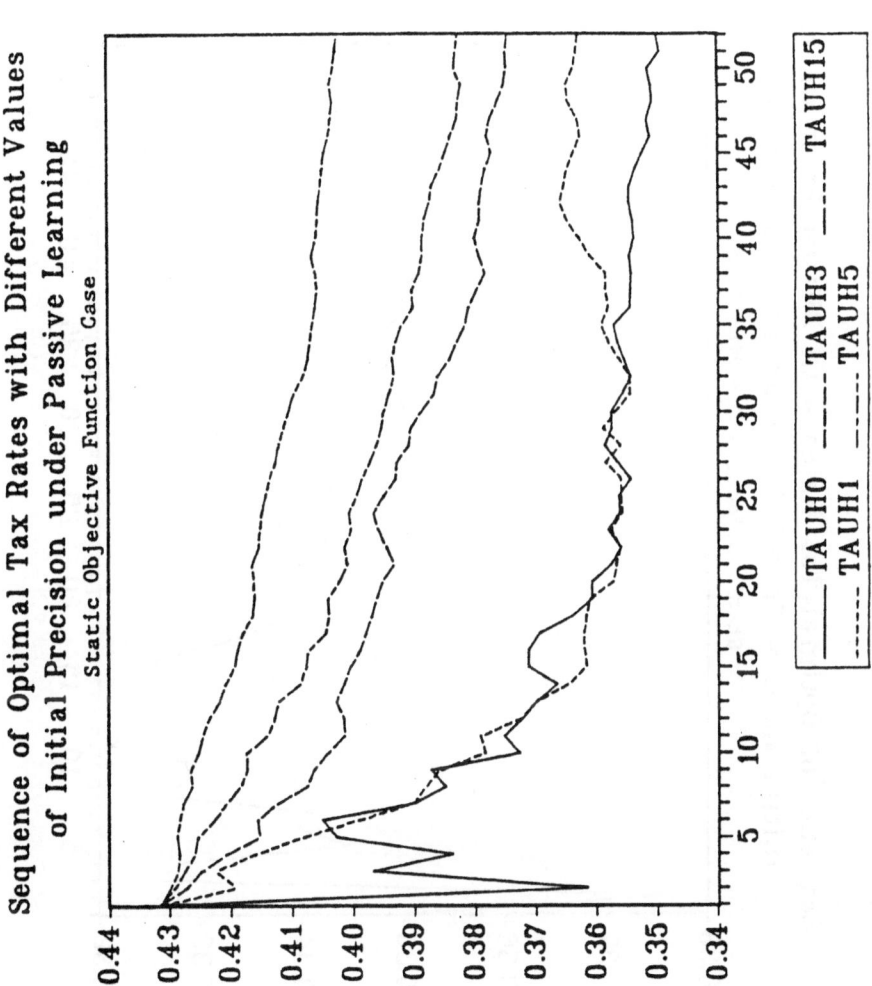

Graph 6

Sequence of Optimal Tax Rates with Diffuse Prior when Initial Mean = True Beta under Passive Learning

Static Objective Function Case

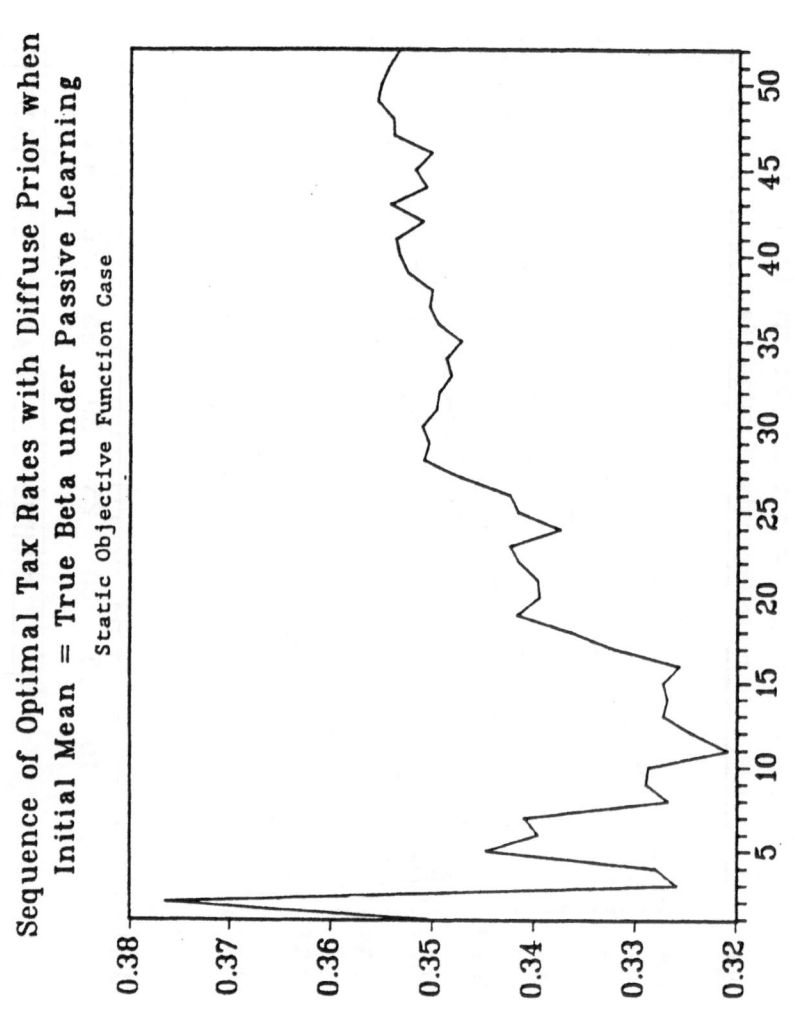

TAUHOBET

Graph 7

Sequence of Optimal Tax Rates with Diffuse Prior under
Passive Learning when Objective Function is Dynamic

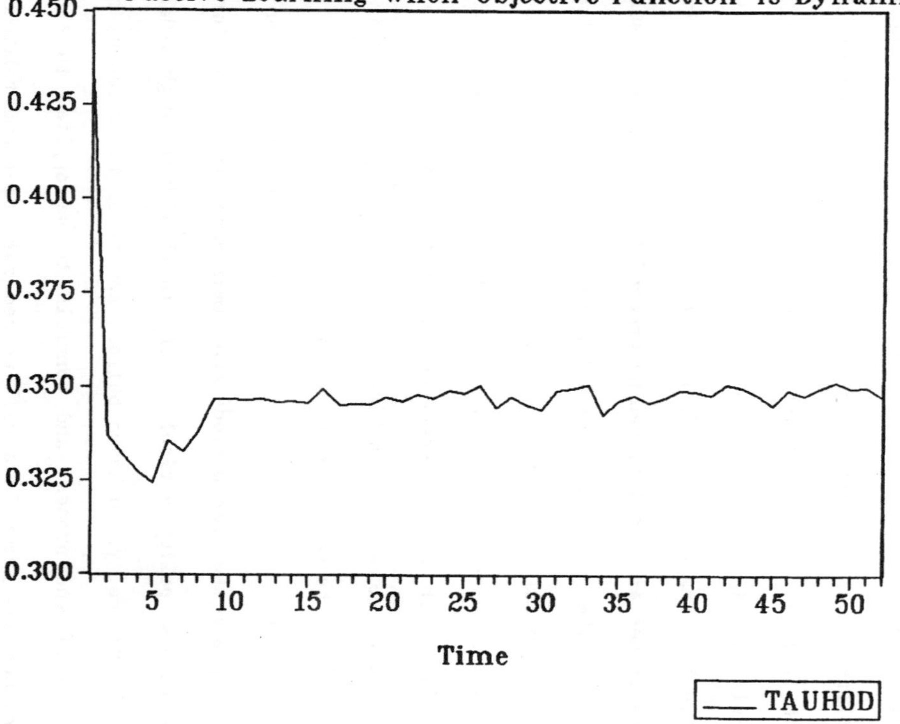

Time

TAUHOD

IV.3 Active Learning

This section has a rich agenda. It first gives the description of the algorithm to compute the active learning r_t sequence. This description involves developing a specific form for the objective function. This form facilitates an argument about the time inconsistency of the policymaker, which follows. It also highlights the impact of the martingale property of beliefs, which necessitates a modification in the objective function for active learning to yield different results from passive learning. Only after these issues are tackled, can we proceed to present the results.

The intuition underlying the notion of active learning is clear but in general complicated to implement: a portion of current payoff is foregone in order to increase the amount of information available in future periods, which in turn enhances optimization in the future. The algorithm utilized to compute the active learning $\{r_t\}_{t=1}^{T}$ sequence in the simulation exercise can be described as follows.

The approach utilized is a sequentialized version of the idea of stacking all the time periods and simultaneously solving for the whole time path of the policy variable - cf. Theil (1964), Hughes-Hallett and Rees, (1983). Future, yet unknown beliefs appear as functions of current beliefs, history of the process and controls to be applied in the future. These functions are known in a well worked out model and taken from an analysis along the lines of chapter II. They follow from the analogues of the update rules (8) and (9). Solving the resulting formulation of the problem therefore accounts for expected gains obtainable from affecting future beliefs via choice of controls, i.e. active learning from the vantage point of the current period. This delivers optimal trade-offs between current payoff and expected future information.

The need for sequential reoptimization in the algorithm arises from the fact that as a period passes, the expectation terms for that period can be replaced by the actual realizations. This changes the amount of information on which optimization for the remaining periods can be based. Hence, for full rationality, i.e. not to ignore available information, a reoptimization becomes necessary since realized beliefs are different from expected beliefs with probability 1 unless there is no more to be learnt. Taking the leading element of the sequences of optimal controls computed this way in each period will yield the sequence of active learning controls - the fully optimal controls, given less than complete information.

The active learning algorithm described above has to be solved numerically since in general it turns out to be analytically intractable. This approach can readily accommodate more complicated objective functions as well, e.g. when there are intertemporal connections other than those arising from the evolution of beliefs or when future beliefs are more richly represented. It can also deal with setups assuming other (conjugate) distributions representing beliefs. For the moment we are concerned with a static period objective function, but at a later stage the objective function will be modified in two respects, one of which is to make it dynamic.

This approach to solving for active learning controls is used not because of the loss of time separability of the objective function in the original dynamic programming problem due to learning. Although this would render dynamic programming suboptimal for the original formulation of the problem, redefining the state variable is a way to get around this, as already mentioned in chapter II. The real reason is that it is computationally much more involved to obtain a dynamic programming solution than to solve for the arguments maximizing the above described form of the objective function, since the former involves iterating in function space. It is not impossible to do so, however: for a simple problem, results of using an algorithm based on an iterated contraction mapping have been reported in Kiefer (1988-89).

As before, $m_1 > 0$ and $h_1 > 0$ is assumed - the latter involves no real loss of generality, since as referred to earlier in this chapter, it is achieved after an initial period has passed. Using (7), (8) and (9) we obtain

$$m_t = \frac{m_1 h_1 + \sum_{i=1}^{t-1} r_i e_i}{h_1 + \sum_{i=1}^{t-1} r_i^2} . \tag{31}$$

Unfortunately, (31) cannot be plugged in as it stands since by (7) it depends on the unknown parameter β and past realizations of the random variable u, neither of which is directly observed. The way out of this problem is to note that by (7') and (29),

$$r_i e_i = \alpha \ r_i - \frac{R_i}{1 - r_i} \qquad \forall \ i < t, \tag{32}$$

a quantity that can be treated as observable for all periods in the past. For the periods when R_i is not yet available, the best the policymaker can do is to plug in the latest mean belief m_t for β and the current expectation of u_i: 0. That is, for these periods (32) becomes

$$r_i e_i - m_t r_i^2 \qquad\qquad \forall \ i \geq t. \qquad\qquad (33)$$

Using this we finally arrive at a version of (29) that conforms to the form of the objective function in Hughes-Hallett and Rees's approach and can thus be utilized in our algorithm. Denoting the present period by s and keeping in mind that as a period has passed, (32) can replace (33) for the corresponding $r_i e_i$ term, the maximand from the vantage point of the present is given by:

$$\sum_{t=s}^{T} \delta^t \ r_t (1 - r_t) \left[\alpha - \left[\frac{m_1 h_1 + \sum_{i=1}^{s-1} r_i e_i}{h_1 + \sum_{i=1}^{s-1} r_i^2} \right] r_t + u_t \right]. \qquad (34)$$

Note that the argument for the necessity of sequential reoptimization is analogous, but not equivalent to the one underlying the time inconsistency phenomenon in macroeconomic policymaking (e.g. Calvo (1978)). Again, as with the Lucas critique, we have a distinct source (learning about the environment by the maximizing agent) of a phenomenon originally derived in a different (game theoretic) setting. In the classic time inconsistency case, it is the fact that another player has already committed his choice of action that changes the constraint of the agent. In our setting, the constraint itself does not change, however its perception by the maximizing agent does, since learning occurs (this is manifested by the fact that (33) can be replaced by (32) above as a period passes and thus becomes history). In either case, a reoptimization yields a different optimal action in general. Thus we have the result that learning is a distinct source of time inconsistency of optimal plans. It is also clear that since only active learning makes explicit plans for future periods based on expectations of future beliefs, this mechanism is only at work for the case of active learning. In the case of passive learning plans for future periods are trivial in the absence of feedforward control and with a static period objective function: they coincide with the controls found optimal for the current period. Since passive learning also changes the perception of the constraint it also implies that after reoptimization the policy variable implemented will differ from the one "planned". Given however that passive learning is not explicitly concerned with the future whereas active learning by its very nature is, it seems fair to restrict usage of the term time inconsistency to the active learning case.

Let us return to the maximand (34). Taking expectation with respect to u in it makes all u_t's vanish. The problem with the setup we obtained is that any attempt at active learning will fail if - as in (34) - the maximand involves only the first moment of the belief distribution. This follows from the martingale property of beliefs, and given our notation can be shown as follows.

Again let s denote the present, j a future period: $j > s$. Let the expectation taken with respect to the distribution of the random noise term u at time s be denoted by $E_s^{(u)}$. Similarly, let $E_s^{(\beta)}$ denote the expectation taken with respect to the belief distribution at time s. Note that the contemporaneous r_i and u_i are independent for all i. This implies that

$$E_s^{(u)} r_i u_i - r_i E_s^{(u)} u_i = 0 \quad \text{for } i = s, s+1, \ldots, j-1.$$

Then from (11), (12) and the above argument,

$$m_j h_j = m_s h_s + \sum_{i=s}^{j-1} r_i [\beta r_i - u_i] \, , \tag{35}$$

and

$$E_s^{(u)} m_j h_j = m_s h_s + \sum_{i=s}^{j-1} r_i [\beta r_i - E_s^{(u)} u_i] \, ,$$

$$= m_s h_s + \beta \sum_{i=s}^{j-1} r_i^2 \, . \tag{36}$$

Applying the operator $E_s^{(\beta)}$ to (36) and utilizing the formula $h_j = h_s + \sum_{i=s}^{j-1} r_i^2$, we get

$$E_s^{(\beta)} E_s^{(u)} m_j h_j = m_s h_s + m_s \sum_{i=s}^{j-1} r_i^2 = m_s h_j \, .$$

Since the left hand side can be rewritten as $h_j \; E_s^{(\beta)} E_s^{(u)} m_j$, we finally obtain upon cancellation of h_j's on both sides:

$$E_s^{(\beta)} E_s^{(u)} m_j = m_s ,$$

the result we were after, a consequence of the martingale property.

In chapter II an argument was given based on the curvature of the value function in beliefs that active learning generically occurs. Now we have the result that in a plausible setup, active learning will collapse to passive learning. Don't these results contradict each other? No, but the second highlights the fact that due to the martingale property the first, more general result will not hold if future beliefs are represented merely by the mean of the future belief distribution. Thus to make active learning nontrivial the specification must be modified to include other characteristics of the future belief distribution apart from its mean. It will turn out that a sensible and simple modification in the discount factor suffices for this. That modification in turn simply follows from assuming the agent to be risk averse in the sense that he prefers more precise to less precise and earlier to later information. Then active learning will lead to a distinct set of control variables.

To approach our goal of specifying the maximand in a form that permits quantification of active and passive learning and at the same time make the specification more general, let us consider two modifications to the original specification of the Laffer curve. One is to create a dynamic constraint by adding a term $E_t \Delta R_t = E_t R_t - R_{t-1}$ to the evasion part of the Laffer curve. Here again E_t denotes $E_t^{(u)} E_t^{(\beta)}$. This by itself will not create room for active learning since still only the means of future beliefs will appear in the maximand. However, periods will have connections other than those via the evolution of beliefs, and this more general specification will provide another benchmark for comparing the simulation results. It will be referred to hereafter as the dynamic objective function case. It allows for the value of the present control variable affecting future payoff since it appears in the weights of various components of the formula for future payoff. However, this effect is conceptually different from (though clearly analogous to) the effect of present controls on future beliefs and through them, on future payoffs - the mechanism underlying active learning.

Including the term $E_t \Delta R_t$ in the evasion term does not interfere with the Laffer restrictions. It also adds realistic features to the maximand. An increase over last period's revenue is valued in itself and a decrease has a negative effect. Increasing present revenue may be tempered by the necessity of future decreases foreseen if the increase is not sustainable. With $E_t \Delta R_t$ present in the maximand it may be reasonable to avoid decreasing present revenue too sharply even if overall maximization dictated controls doing that in the absence of this term. Thus in a sense, inertia in changing the magnitude of r is introduced - a property argued to add a realistic feature to the model.

The second proposed modification in the objective function is the one actually aimed at facilitating active learning. It relies on the argument that the maximizing agent should be risk averse in the sense of valuing more highly information which is more precise and preferring precision in information on β to be obtained earlier rather than later. A simple method of capturing this is to incorporate precision into the period maximand multiplicatively. This implies that it will be merged with the discount factor δ^t. Taking into account the additional requirement that the composite discount factor thus obtained should still converge to zero, the following specification is suggested:

$$\Delta_t = \delta^t h_t^{\nu/t} . \tag{37}$$

In (37), the parameter ν is a positive constant reflecting the degree to which precision is preferred by the agent. A higher ν implies that precision is valued more. In this sense, the value of ν reflects the degree of risk aversion. The parameter δ reflects the rate of time preference, and h_t, precision is given by (8). The composite factor Δ_t is positive and can be shown to approach 0 as t approaches infinity. Since Δ_t explicitly involves the attained precision of the belief distribution, nontrivial active learning will be possible with the maximand at any time s defined as a constant plus

$$\sum_{t=s}^{T} \Delta_t \ r_t (1-r_t) \left[\alpha - \left[\frac{m_1 h_1 + \sum_{i=1}^{s-1} r_i e_i}{h_1 + \sum_{i=1}^{s-1} r_i^2} \right] r_t \right]. \qquad (38)$$

Note that all the results of chapter III remain valid with this change in the specification of the objective function. First, because all they relied on was that learning occurs (not necessarily active learning). Second, and more importantly, only the discount factor has been modified, not the specification of the Laffer curve. The presence of a time varying discount factor in no way overturns the arguments given in chapter III.

To avert the criticism that (37) is arbitrary, an effort is made to base the choice of this time varying discount factor on sound arguments in terms of fundamentals. Let the policymaker possess the period utility function (already a derivative one, in terms of revenue, not in terms of social welfare) given by

$$U(R(r_t)) = \frac{\log R(r_t)}{\epsilon_t^2} \qquad (39)$$

where $\epsilon_t = \beta - E_t(\beta)$. This error is not observable by the policymaker but the resulting formulation will justify its inclusion. The utility function (39) conforms to Arrow's requirements for a reasonable representation of risk aversion: it yields an Arrow - Pratt measure of absolute risk aversion of $1/R$, and of relative risk aversion of 1. Hence it implies decreasing absolute risk aversion (DARA) and constant relative risk aversion (CRRA) deemed appropriate in general in Arrow (1965). Maximizing its expectation yields a period maximand having a multiplicative $[E(\epsilon_t^{-2})]$ term. Replacing it by $h_t = (E(\epsilon_t^2))^{-1}$ means (by Jensen's inequality) mitigating the effect on utility somewhat, but the direction of the impact this term has is clearly unchanged. Adding a parameter to achieve flexibility in setting the rate of risk aversion, transforming by taking the t-th root and merging with the original discount rate yields the time varying discount factor (37) proposed above.

To summarize: in the simulation exercise (38) is utilized in computing the active learning sequence of controls with the term $r_i e_i$ substituted out by using (32) or (33) depending on whether i is a period in the past or not. The dynamic version of the objective function contained the additional $E_t \Delta R_t$ term in the evasion part. Note that for any given time period s the problem is completely deterministic but highly nonlinear. Randomness still has a role to play: it affects the value of the constant in the above maximand, corresponding to the past periods as well as the path of mean beliefs and precisions. As mentioned before, the "rolling substitution" of (33) by (32) as time evolves is a source of time inconsistency distinct from the usual one. It is also necessary for fully optimal behavior since not performing it would amount to not using the latest available information. Hence with active learning, rationality in the above sense necessarily raises the possibility of time inconsistency of optimal policy plans.

The method utilized to maximize the objective function was the Broyden-Fletcher-Goldfarb-Shanno algorithm, a variant of the Newton-Raphson method. (For the properties of the Broyden-Fletcher-Goldfarb-Shanno algorithm, see Dennis and Schnabel (1983), for a discussion of the Newton-Raphson method and related algorithms, cf. Amemiya (1985).) As before, the values of the parameters were set as follows: $\alpha = 1000$, $\beta = 900$,

s^2 - 15000, ν - 15. The term diffuse prior refers to h_1 - 0, nondiffuse prior to h_1 - 3.

Before proceeding to the results obtained, an overview of the various setups in the computer experiments is given. The maximand from the vantage point of period s is denoted by M_s, $1 \leq s \leq T$.

Table 1.　　　　Overview of Maximands

1. Static Objective Function

　　1.a. Passive Learning

$$M_s - E_s \sum_{t-s}^{T} \Delta_t \tau_t (1-\tau_t)[\alpha - \beta\tau_t + u_t]$$

　　1.b. Active Learning

2. Dynamic Objective Function

　　2.a. Passive Learning

$$M_s - E_s \sum_{t-s}^{T} \Delta_t \tau_t (1-\tau_t)[\alpha - \beta\tau_t + E_t \Delta R_t + u_t]$$

　　2.b. Active Learning

IV.3.1 Results - Active Learning

Active learning implies a different set of chosen tax rates if the specification is modified to include the time varying discount factor (37). The deviation from the passive learning path of tax rates is directly related to the degree of risk aversion the policymaker is assumed to have. Allowing for active learning in most cases significantly increased the attainable value of the objective function. Allowing for dynamics other than via the evolution of beliefs also had an impact on beliefs and a significant one on the controls chosen (the value of the objective function attained has of course changed). It also increased the computation time needed to obtain a solution, since intertemporal linkages interacted in a complicated way.

While the degree of nonlinearity of the objective function certainly raised doubts on whether the solution paths would be unique, test runs with several reasonable (i.e. above 0.3) initial values resulted in practically identical solution paths. This uniqueness property is somewhat surprising given the highly nonlinear nature of the maximand. It may be due to the $r_t(1-r_t)$ term, which implies that any maximum is restricted to be in the $(0,1)$ interval. Given an initial value for r_1 of above .3, this interval seems usually to contain only one optimal value for r. In the absence of this term or one implying a similar strong restriction on the range of the optimal controls, the problem of nonuniqueness of the solution is sure to arise. Problems did arise in the simulations however when extreme initial values were assigned to r_1. For example when initial values were set at 0.01, the maximization algorithm wandered outside the admissible region $(0,1)$ for r and bogged down. For r_1 around 0.2 the algorithm sometimes bogged down and sometimes yielded reasonable results after dramatically more iterations than for the case $r_1 > 0.3$. This may be either a problem of the particular algorithm employed or of the particular values plugged into the objective function and not necessarily one with the theoretical underpinnings of the simulations.

No explicit terminal condition has been imposed on the solution path. Since the assumption of a known, finite terminal period is in many contexts not tenable however, the following implicit terminal condition imposing smoothness on the path of controls was employed. The value of T has been set to be beyond the true (or anticipated) terminal period T^* and results were only considered for periods $1,2,...,T^*$. This approach resulted in smooth paths of control variables in all cases tried. Although there was no indication that breaks would have occurred in the absence of this precaution, it cannot be claimed that it was unnecessary since by any reasonable measure convergence of beliefs occurred by the terminal period in most cases tried and this ensured smoothness of the controls anyway.

A distinctive feature of the resulting time paths of control variables for higher values of the risk aversion parameter ν was that the initial couple of controls (and hence except for some settings of the parameters, resulting revenues) were so different from the subsequent ones that they could plausibly be labelled outliers. This phenomenon would be even stronger in models where optimal actions are more sensitive to changes in beliefs. Thus learning, in particular active learning, is a plausible mechanism for "endogenous" outlier generation. (For a similar result in a two-armed bandit setting, consult the marketing example given in Kiefer (1988-89).) This mechanism is not due to some external effect completely unrelated to the system - like a keypunching error at data entry. Rather, it is a consequence of an assumption made about the nature of the data generating mechanism itself, and a reasonable one, too: the assumption that agents initially have significantly less than complete information about their environment. This of course does not necessarily make life easier for the econometrician using the data. It certainly highlights the fact however, that employing a dummy variable whenever an outlier is detected is not necessarily optimal.

The shape of the time paths of control variables for all parameter constellations and specifications implying significant learning effects was always very similar: a decaying additional component was superimposed on an otherwise stationary path. This fits in nicely with the discussion in section III.2.4 of learning being a distinct source of nonstationarity. While this shape is not universal, since it need not occur for situations where the bigger is better property does not hold, for the class where it does hold, it can be called the typical shape. In a more general context, learning superimposes a decaying effect but this effect is not necessarily always of a uniform sign.

Initial mean beliefs were again quite volatile from one period to another depending more on initial precision and the variance of the noise variable than on how close they initially were to β. The effects of this on the optimal tax rates and hence on the value attained by the objective function are somewhat masked by the form of the period maximand: it is still dominated by the term $r_t(1-r_t)$.

Precision behaved very much as expected, it increased at a declining rate, more quickly when a premium was placed on it in the specification designed to allow room for active learning, especially when the risk aversion parameter was higher.

A rough idea of how important experimentation is can be obtained by comparing the sequence of the bias terms: $m_t - \beta$, the values attained by the objective function with active and passive learning, or by comparing the paths of the tax rates or revenues under the two learning schemes. With any of these methods, experimentation is naturally seen to gain importance only with the more general objective function that gets around the martingale property and to vary directly with the rate of risk aversion of the policymaker. There clearly are situations in which experimentation (or probing) is insignificant, but also situations in which it plays a major role. The simulation results thus tend to confirm the claim of chapter II that with a sufficiently general objective function active learning has a more than negligible role to play. Moreover, a sufficient set of conditions i.e. specific constellations of parameter values for this to happen have been identified. The results show that the exact way this role is measured, is very important however. Optimal decisions may not change much while mean beliefs jump around dramatically. If the importance of experimentation is measured by comparing revenues, care must be taken that it is not the effect of the noise term that shows up in our measure.

This chapter concludes with graphs describing results of the simulations. The way chosen to present the results is slightly different from the usual method of presentation involving summary statistics and averages over a large number of realizations. The reason for avoiding averaging over realizations is that the resulting average sequence of tax rates would not correspond to any valid path of beliefs, in particular it would not correspond to the average path of beliefs that would be presented next to it. As already mentioned, different realizations were quite similar, so the realizations displayed can be argued to be representative of the sample of realization paths generated in the simulation exercise.

On comparing graphs 8 and 9 it is obvious that the static and dynamic objective function yield results that differ much more than was the case for passive learning. Mean beliefs converge towards the true β value again and obviously a higher initial precision has the same smoothing effect on the path of mean beliefs (and hence on optimal tax rates) as before. Note that the agent does a markedly better job of approaching the true β value with the dynamic objective function than with the static one if initial beliefs are seriously mis-specified: the bias $|m_t - \beta|$ is significantly and uniformly smaller than with the static objective function on graph 9. Thus dynamics other than via the evolution of beliefs can make a difference with active learning - a rather intuitive result. The fact that convergence is somewhat slow in general is a consequence of the large variance of the noise term.

Graph 10 illustrates the time inconsistency phenomenon. It presents the plans for the optimal tax rate to be applied in the final period as computed from the vantage point of the period shown on the time axis for the four cases arising as combinations of dynamic or static objective function and diffuse or nondiffuse prior. The extent of revisions i.e. the degree of time inconsistency is greatest in the initial periods - this is obviously a reflection of the fact that the largest adjustments in beliefs also occur in these periods. Applying a nondiffuse prior made the extent to which optimal plans were revised from period to period smaller for both types of objective functions. The static objective function with diffuse prior yielded a time inconsistency profile very similar to these. The dynamic objective function with a diffuse prior produced a markedly different result however: time inconsistency was significant and persisted for much longer than for the other cases. This setting is therefore a concrete example showing that the time inconsistency phenomenon induced by learning can be of a significant magnitude. It is also interesting to note that even after 50 periods the planned optimal tax rate of this setting is different from that of the other settings which already clustered together by the 20th period.

Graph 11 displays the sequence of actually applied tax rates with the dynamic and static objective function when prior precision was 0. The values are the leading elements of the whole vector of tax rates found to be optimal for the remaining time horizon given updated current beliefs for each time period. It is not surprising that the dynamic objective function allowing for more complex intertemporal interactions results in a more variable time path. Two properties of this path are remarkable however. First that it differs significantly from the corresponding time path generated using the static objective function throughout the whole time segment considered. Second, that the effect of intrinsic dynamics introduced by the $E_t \Delta R_t$ term has overwhelmed the "bigger is better" effect that would have called for convergence of applied tax rates to the optimal level corresponding to the true β value from above (to generate higher precision in the early precision). This result persisted for a higher setting of ν as well. Hence we have uncovered another mechanism that can invalidate the bigger is better result described in Appendix B: intrinsic dynamics in the objective function.

Graph 12 displays the optimal active learning tax rates corresponding to the static and objective function again, but this time for the case of a nondiffuse prior. As always, mis-specified beliefs held confidently resulted in a smoother path of tax rates. The dynamic objective function again yielded somewhat more variable tax rates.

Graph 8

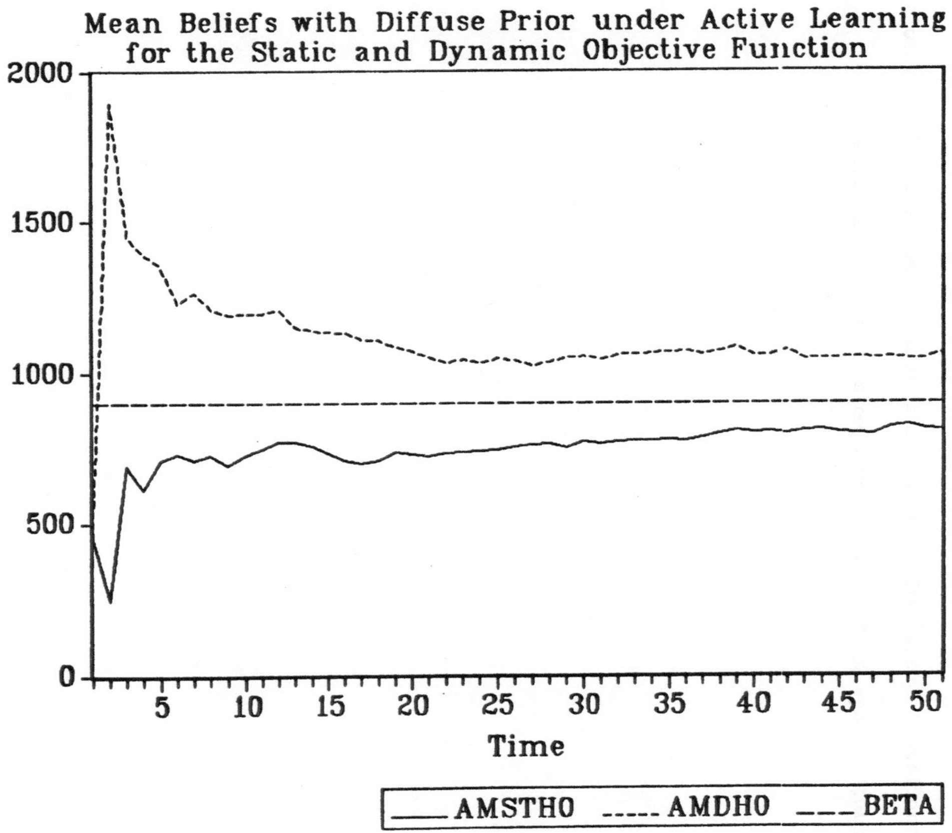

Mean Beliefs with Diffuse Prior under Active Learning
for the Static and Dynamic Objective Function
2000
1500
1000
500
0
5
10
15
20
25
30
35
40
45
50
Time
AMSTHO
AMDHO
BETA

Graph 9

Mean Beliefs with Nondiffuse Prior under Active Learning
for the Static and Dynamic Objective Function

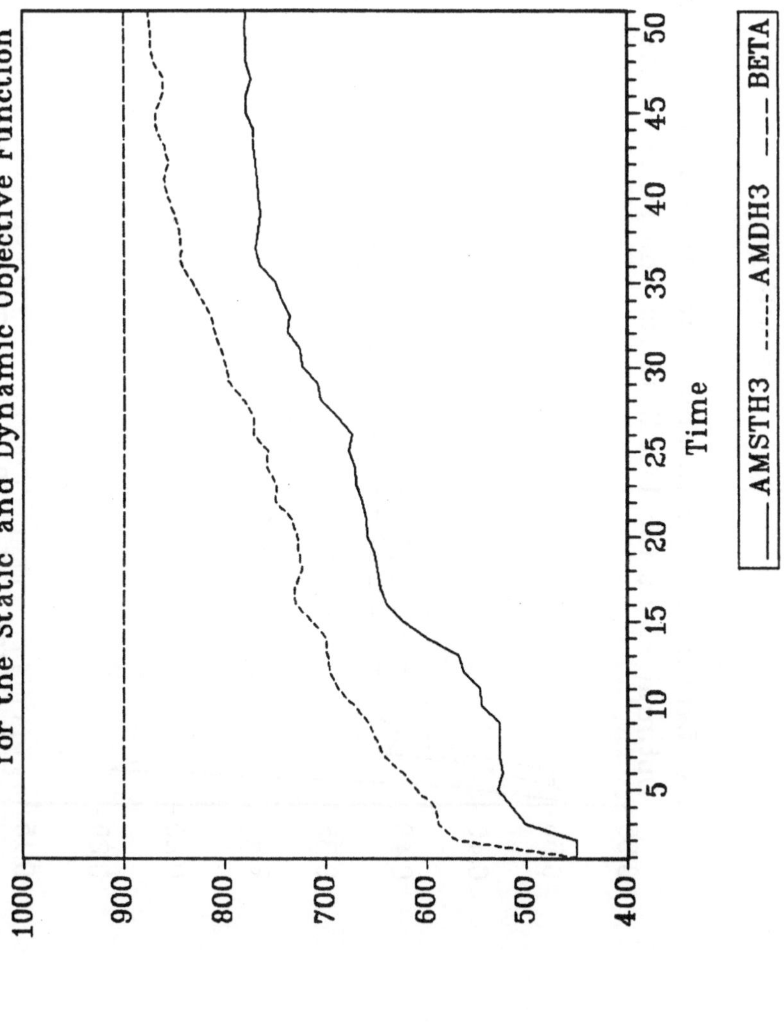

Graph 10

The Extent of Time Inconsistency in Various Settings:
Optimal Plans for Tax Rate for the Final Period

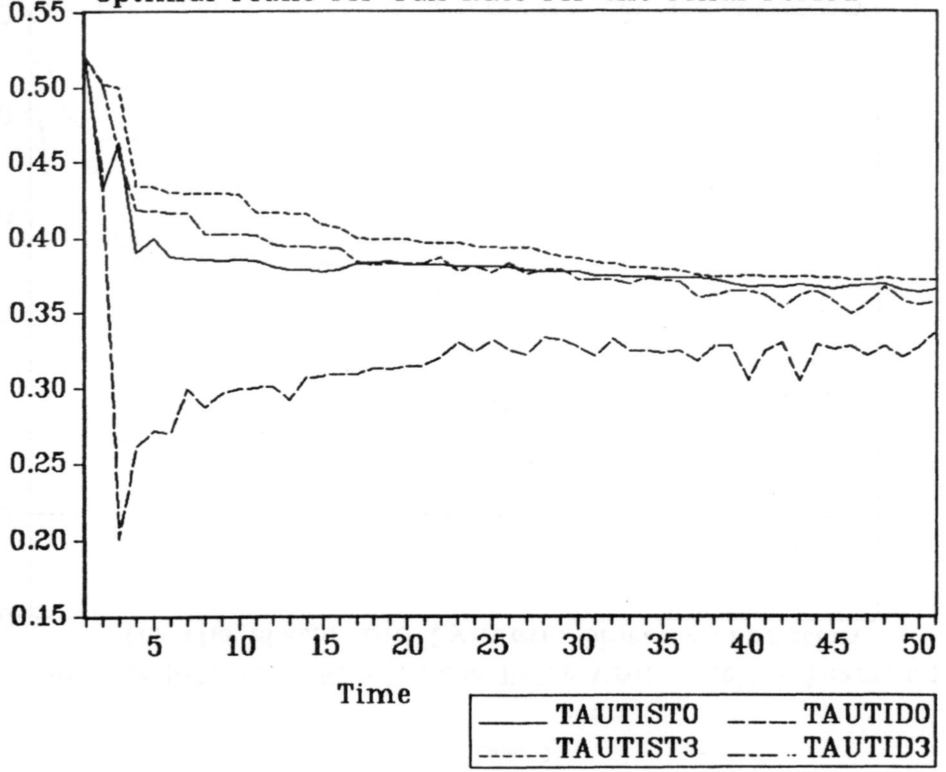

Time

TAUTIST0 — TAUTID0
TAUTIST3 — TAUTID3

Graph 11

Sequence of Applied Optimal Tax Rates under Active Learning for Static and Dynamic Objective Function

Diffuse Prior Case

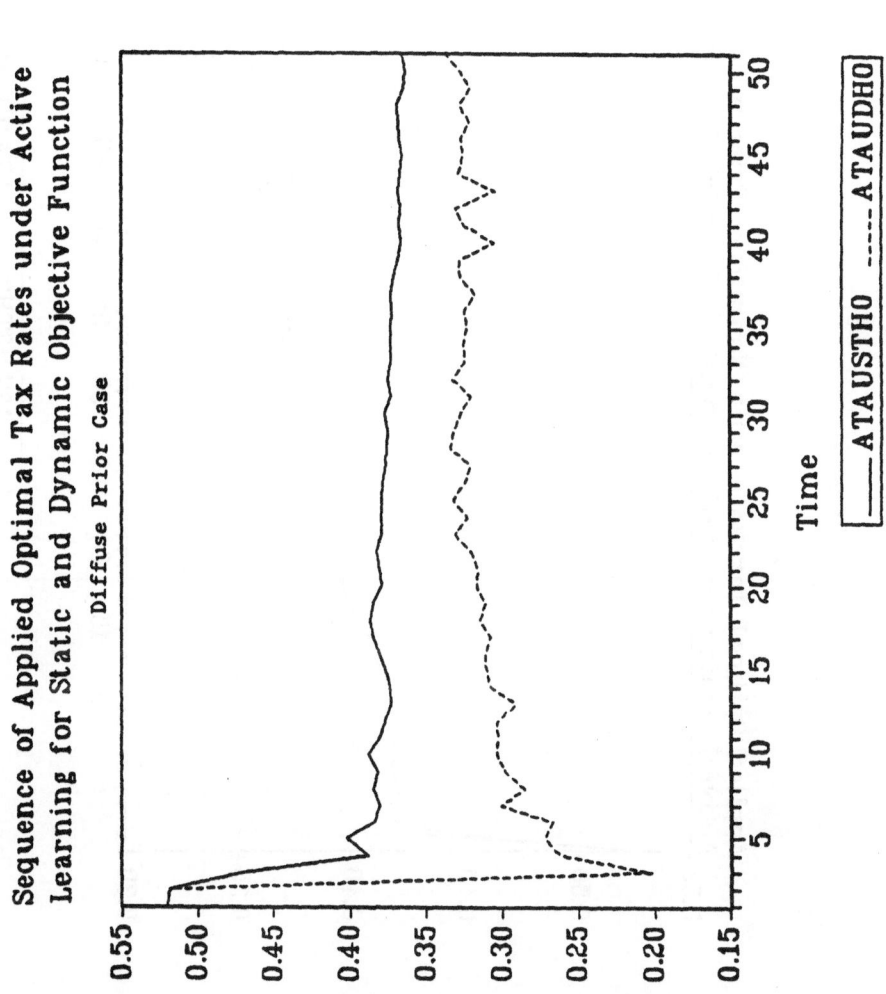

Graph 12

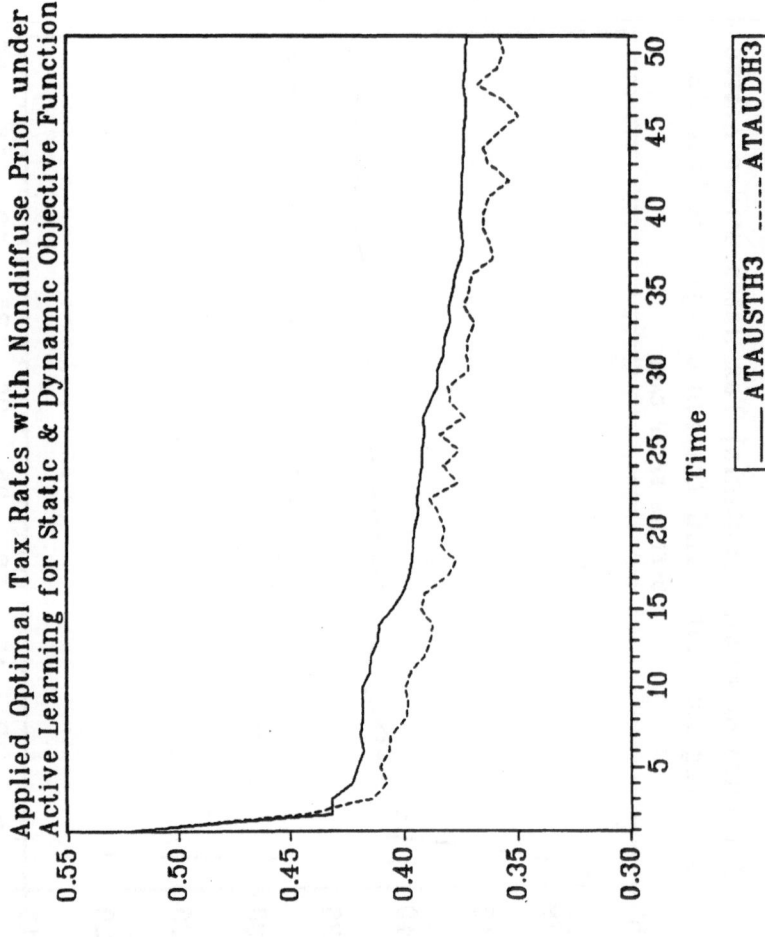

Applied Optimal Tax Rates with Nondiffuse Prior under Active Learning for Static & Dynamic Objective Function

V. Tests for Exogeneity

V.1 Overview

Engle, Hendry and Richard (1983) and Hendry and Richard (1983)
argue in a convincing manner that their notion of exogeneity is the natural
one to use instead of classical (strict) exogeneity if the goal of the
econometrician is to use the simplest possible model without loss of
relevant statistical information. Geweke (1984) on the other hand reveals
the opposite side of the coin: while it is weak exogeneity that we are
really after, that concept by itself cannot generate refutable hypotheses
(see also Basmann (1965)). In other words, a model can always be constructed
and parameters of interest can be chosen for which a designated set of
variables is weakly exogenous - a simple way to do this is to specify the
conditional model and the marginal model independently. Therefore weak
exogeneity by itself does not supply enough restrictions that it can be
subjected to statistical tests. So, in order to make weak exogeneity
operational, concepts that generate empirically refutable hypotheses and
which under reasonable assumptions are compatible with weak exogeneity are
needed. And here is where we come a full circle: strict exogeneity is such a
concept - though not the only one. Testing for strict exogeneity is possible
(Sims (1972), Williams, Goodhart and Gowland (1976), Ciccolo (1978)), and
such tests can be interpreted as joint tests of strict exogeneity and the
assumptions that link weak exogeneity to strict exogeneity.

This is the best one can do in testing for weak exogeneity if the exact specification of the marginal model is to be avoided. The fact however that the exogeneity status of a variable is subjected to testing is tantamount to saying that its generating process is suspected to be irrelevant. To specify a marginal model just to find out whether we need to specify it is not very appealing. Also, the correctness of the answer we get hinges on whether the specification of the marginal model is correct. In summary, if one follows this approach, the only way to prove definitively that specification of the marginal model is not necessary is to find the correct specification for it. In this case however, whether or not it is necessary, we already have it, and we can't lose if we use it (computer constraints permitting). Engle (1984) contains Lagrange - multiplier based tests of weak exogeneity of this sort.

Returning to tests based on the conditional model only, in the case of strong exogeneity one of its ingredients, Granger noncausality is directly testable. In the case of super exogeneity, if a sample of sufficient size is available that covers two (or more) regimes, a Chow test (cf. Chow (1960)) can directly test for the structural invariance component. This only works if the sample period covered all regimes considered, however. Therefore it does not shed light on the question of structural invariance if policy experiments involve venturing to "uncharted waters", i.e. cases that do not lend themselves to a Chow test due to insufficient data in the sample.

Weak exogeneity is a common element of all three Engle, Hendry and Richard exogeneity concepts. Thus it should be subjected to testing if any of the notions is treated as a hypothesis, rather than a maintained assumption for a particular model at hand. To test weak exogeneity it was argued to be necessary to link it to a testable concept such as strict exogeneity. This in turn can be tested by testing for unidirectional Granger causality from the exogenous to the endogenous variable (cf. Geweke (1984)).

We then have the result that tests for Granger causality are useful both as direct tests for strong exogeneity and strict exogeneity and as indirect tests for weak (and hence strong and super) exogeneity. To sum up, suitable versions of Granger causality tests can be interpreted as at least indirect tests for each of the Engle, Hendry and Richard exogeneity concepts. This is the reason for considering only Granger causality tests in this chapter.

The assumptions made in Sims (1972) and by those following his lead when testing for unidirectional Granger causality involved stationarity and lack of deterministic components in the time series utilized. Stationarity does not hold for the time series generated by a learning agent - this has been shown in chapter III. Given this, it would seem that the presence of learning by itself presents a formidable problem for the empirical econometrician. The same phenomenon that generates loss of exogeneity on all counts for a policy variable, also generates nonstationarity of the time series involved. To make things worse, this nonstationarity is not always of a nature that can be taken care of using a simple transformation such as taking logarithms or de-trending. Depending on the exact set of assumptions made on the information level and information processing mechanism of the economic agent, it can be a smoothly decaying component added to an otherwise stationary process as in the case of passive Bayesian learning, or it can be more complicated. The latter may occur when the support of the belief distribution is truncated as a part of the learning mechanism, when that distribution is non-normal, or when learning is active or not Bayesian. The path of the resulting time series is also affected by the exact nature of the objective function of the economic agent: whether it incorporates higher moments of the belief distribution and with what weights. This will depend among other things on the rate of risk aversion of the agent. The realizations of the noise variable also affect the path - not only in the usual sense of contemporaneously shocking the dependent variable, but also by shocking beliefs away from the truth if too large and cushioning the effect of active learning if negative (to use the example of a case when the bigger is better result holds).

Let us now examine the possibility that the time series contain deterministic components or are nonstationary. These cases can be treated together following Hosoya (1977). Deterministic components can cause nonstationarity of the observed time series. If this is not the case, then the deterministic component is just a constant, easily taken care of. Consequently it is enough to concentrate on nonstationarity whether or not caused by deterministic components. If nonstationarity is "removed" by the econometrician, to obtain data conforming to requirements of the tests like those employed in Sims (1972) and Ciccolo (1978), an additional source of error in the data is introduced, since as was argued, even a simple learning model may imply extremely complicated nonstationarity properties for the data. If nonstationarity is not removed, Granger causality tests are still available but may have considerably less power against some alternatives - cf. Geweke (1984).

Having highlighted some of the caveats facing the empirical econometrician now we shall go over the exact formulation to be used in testing for each of the exogeneity concepts. A view of how the obtained results should be interpreted will be offered.

The chapter will conclude by reporting the results of the proposed tests using the data generated in the simulation exercise reported in the previous chapter and contrasting the findings with what was expected on the basis of the theoretical arguments in chapters III and IV.

V.2 Formulation of the Exogeneity Tests

As argued above, tests of various notions of exogeneity can all be related to tests for unidirectional Granger causality from endogenous to exogenous variables. As in chapter III, rather than stipulating a specific model, we take the more general approach of assuming only that the investigator hypothesizes an econometric model with R an endogenous and s an exogenous variable. For sufficient generality, assume that this model is a complete linear dynamic simultaneous equation model. Since Granger causality tests aimed at testing the null hypothesis that s is strictly exogenous merely confirm the list of exogenous variables in a model if the null hypothesis is accepted but reject even this loose formulation otherwise, this approach ties in neatly with what tests of this sort can and cannot achieve.

The assumptions needed for implementing the tests are the following. The relevant universe of information at time t consists of the history $\{s_s\}$ and $\{R_s\}$, $s < t$. The criterion for comparing the goodness of alternative forecasts is mean squared error. Only linear predictors are considered. The earlier literature following Sims (1972) also assumed that the time series R and r are (or at least can be transformed to) stationary, purely nondeterministic time series with autoregressive representations. The assumptions on stationarity and purely nondeterministic nature of the time series is not essential for our purposes: they can be given up (Hosoya (1977), albeit at some (not precisely known) loss in the power of the tests, as already argued above. In the tests reported here no corrections were applied for nonstationarity and deterministic components in the data. Moreover, the tests have only asymptotic justification.

After defining the Granger and Sims variants of the test, the Wald version of each is chosen to be implemented, since that was shown to have the most power both asymptotically (this result is based on comparison of "approximate slopes" analogous in nature to the asymptotic power of tests) and empirically in small samples (based on Monte Carlo experiments) in Geweke (1984). All tests utilized are consistent.

Let the null hypothesis for the Granger variant be

H_0^G : r is strictly exogenous in the model with R endogenous.

Consider also the formulation

$\tilde{H}_0^G$: $b_{2s} = 0$ $\forall s$,

where $\tilde{H}_0^G$ refers to (41) below. Testing H_0^G given our assumptions can be achieved by testing whether the two regressions

$$r_t = \sum_{s=1}^{p} a_{1s} r_{t-s} + \epsilon_{1t} \tag{40}$$

and

$$r_t = \sum_{s=1}^{p} a_{2s} r_{t-s} + \sum_{s=1}^{q} b_{2s} R_{t-s} + \epsilon_{2t} \tag{41}$$

are equivalent, with $\tilde{H}_0^G$ stating that they are. This test is labelled the Granger variant, since it obviously reflects Granger's definition of causality directly by asking whether the inclusion of lagged endogenous variables improves the prediction of r_t. Under H_0^G it doesn't and if this null hypothesis is accepted then we have that R does not Granger cause r while the hypothesized model of course implies that r Granger causes R. This state of matters coincides with the Koopmans view of an exogenous variable as one that affects but is unaffected by the endogenous variable, of which strict exogeneity is a statistical formalization. Hence we have at hand a test for strict exogeneity of r.

The Sims version relies on a theorem in Sims (1972) in effect restricting the coefficients of future exogenous variables to be zero in a regression with the endogenous variable as left hand side variable and lagged endogenous, past, current and future values of the exogenous variable as right hand side variables if H_0^S : R does not Granger cause r is to be true. As with the Granger variant, truncation at ingeniously or arbitrarily chosen points is necessary with a finite data set and thus the test boils down to a test of the hypothesis $\tilde{H}_0^S$: $d_{2s} = 0$ $\forall s < 0$ in (43), the second of the following pair of equations:

$$R_t = \sum_{s=1}^{p} c_{1s} R_{t-s} + \sum_{s=1}^{m} d_{1s} r_{t-s} + w_{1t} \tag{42}$$

$$R_t = \sum_{s=1}^{p} c_{2s} R_{t-s} + \sum_{s=-r}^{m} d_{2s} r_{t-s} + w_{2t}. \tag{43}$$

Evidently this formalization reflects the Sims restrictions described above. In both cases, serial correlation in the ϵ_j and w_j series ($j=1,2$) must be checked for, since the Granger definition and the Sims theorem are both only valid with infinite summations and therefore truncation points may be such that significant lags are excluded which would result in serial correlation of the error terms. Serial correlation should not be "corrected for" by a transformation since its cause is now known, rather, it should be reduced by increasing the lag and lead lengths in (40) to (43). This of course must be traded off against the decrease in degrees of freedom, but this is not a serious concern for a generated dataset, since additional data can always be generated. Somewhat more important is the issue of nominal versus real significance levels if the optimal leads and lags in the implementation of the tests has been established via trial and error. If the null hypothesis of r being exogenous is rejected solely as a result of learning behavior by the policymaker for the case when the econometrician happened to pick optimal leads and lags in his empirical tests, then evidence is obtained that the question posed in this study is an interesting one with far reaching consequences. The weight that this evidence carries however is diminished somewhat by the fact that ideal circumstances have been assumed for detecting the effect of learning. On the other hand, the finding that even under ideal circumstances the effect is not detectable would be a strong (though negative and extremely worrisome) result.

Let us now turn to defining the test statistics following Geweke (1984). Let the estimated variance calculated from ordinary least squares residuals in the regression (40) to (43) be $\hat{\sigma}_i^2$, i=1, 2, 3, 4, respectively. Thus:

$$\hat{\sigma}_j^2 = \frac{1}{T} \sum_{t=1}^{T} \hat{\epsilon}_{jt}^2 \qquad\qquad j=1,2 \qquad\qquad (44)$$

and

$$\hat{\sigma}_3^2 = \frac{1}{T} \sum_{t=1}^{T} \hat{w}_{1t}^2 \ , \qquad\qquad \hat{\sigma}_4^2 = \frac{1}{T} \sum_{t=1}^{T} \hat{w}_{2t}^2 \qquad\qquad (45)$$

Let superscript G and S stand for Granger and Sims variant, respectively. Then the test statistic for the Granger variant of the test is given by

$$D_T^G = T \left[\frac{\hat{\sigma}_1^2}{\hat{\sigma}_2^2} - 1 \right] \qquad\qquad (46)$$

and for the Sims variant by

$$D_T^S = T \left[\frac{\hat{\sigma}_3^2}{\hat{\sigma}_4^2} - 1 \right] , \qquad\qquad (47)$$

both of which are asymptotically distributed as χ^2 with degrees of freedom q and r respectively.

The Sims version of the test introduced here differs from the one in Sims (1972) in two respects: first, there an F test was used, second, Sims constrained c_{js} in (42) and (43) to be identically 0, j=1,2.

Now, some remarks. First, the Granger and Sims versions of these tests are not related in any functional way for finite samples. Therefore, it is meaningful to perform and report the results of both tests, since one may reject the null hypothesis while the other does not using the same set of data. Also, in both versions the tests, as formulated, ignore instantaneous causality. The results therefore should be interpreted keeping this in mind. Geweke (1984) contains the statistical apparatus for tests allowing for instantaneous causality as well. Finally, Granger noncausality from R to r is a necessary but not sufficient condition for r to be strictly exogenous in a specific econometric model. Consequently, the result of these tests can refute but not decisively accept the null hypothesis of r being strictly exogenous in the specified econometric model.[1]

We are now equipped with the statistical apparatus for testing for strict exogeneity of a variable in a class of models. The next step is to establish the link allowing us to interpret the results of these tests as supplying evidence for a joint hypothesis implying weak exogeneity of the same variable.

The link we are after is a condition to guarantee that given strict exogeneity of a variable in the conditional model, it can be treated as fixed for the purpose of inference without loss of relevant sample information. For this to be meaningful, the parameters of interest need to be specified: assume that they are a nonsingular transformation of the parameters of the conditional model and they include all parameters in that model necessary for efficient estimation - cf. section II.2.1. Note that hereby all parameters of the marginal model have been assumed to be nuisance parameters. This assumption together with strict exogeneity of the variables under scrutiny ensures consistent estimation of the parameters of interest using the conditional model only. If this estimation is also efficient then we have that the strictly exogenous variables are also weakly exogenous for the estimation of the parameters of interest. The key issue here is whether or not to achieve this one has to explicitly require that requirement (a) hold. This can only be determined if the concrete model is known. If requirement (a) need not be imposed and some other testable restriction (R, say) suffices for the concrete model at hand, then strict exogeneity and (R) imply weak exogeneity, making it indirectly testable by testing for the joint hypothesis H_0^J: r is strictly exogenous and (R) holds. Parameter restrictions which can serve as the restriction (R) can be found in Example 3.2 in Engle, Hendry and Richard (1983) for a specific model. The test described above is indirect since it does not rely on requirement (a) or (b). If on the other hand, both requirement (a) and (b) have to be postulated by the econometrician to ensure efficiency of estimation using the conditional model only, then weak exogeneity is not testable. It then becomes a maintained hypothesis. In this case, only strict and strong exogeneity (and possibly super exogeneity, under the circumstances explained earlier) is testable without explicitly specifying the marginal model for the exogenous variables.

It has been shown how the test result for strict exogeneity can be augmented with additional restrictions to yield an indirect test for weak exogeneity. We now have a test that the model of the econometrician will have to fail for the theoretical results of chapter III to be substantiated. If weak exogeneity can be rejected by the econometrician, then learning has been theoretically shown to cause an empirically detectable breakdown of the usually maintained hypothesis of policy variables being exogenous in all of the following senses: strict, weak, strong and super exogeneity. For the cases of strict and strong exogeneity the tests employed are direct while for the other concepts they are indirect in the sense already explained. If strict exogeneity is not rejected, then the specification of the model with the policy variable exogenous is weakly supported in the sense that a refutable hypothesis of this specification failed to be rejected.

V.3 Results of the Exogeneity Tests

In what follows it will be assumed that some additional restriction of the type (R) was found and also accepted using an appropriate test. Therefore results presented below are to be interpreted as indirectly testing for weak exogeneity as well (having appropriately modified the significance levels). The results of the tests are summarized in the following tables. The lag and lead lengths have all been chosen to be the same value and are always reported in the tables as lag length. The null hypothesis is always H_0: R does not Granger cause r. The critical values of the tests with $r=q=5$ are:

$$11.07 \quad \text{at the .05 significance level}$$
$$\text{and} \quad 15.09 \quad \text{at the .01 significance level.}$$

For interpretation of the results to be possible, the variable names have to be explained. In Table 2 names of dependent variables are displayed in the "H_0 accepted" or "H_0 rejected" column according to the outcome of the test based on regressions (40) and (41) with $p=q=5$. The particular setup is described in the second part of the variable name, appearing after "tau". The first five entries correspond to the passive learning, static objective function experiments reported in Graph 3, with initial precisions taking the values 0, 1, 3, 5, 15 respectively, as it appears in the variable names. Taudph0 stands for the optimal tax rates with dynamic objective function, passive learning and zero initial precision. The rest are active learning situations: if a "d" occurs in the variable name, then the dynamic objective function was applied, if "s", then static. Finally, again, h0 and h3 stand for initial precision equalling 0 and 3, respectively.

Table 2.	Exogeneity Test		
Granger Variant		Lag Length - 5	
Reject H_0 when LHS variable is	Value of test statistic	Accept H_0 when LHS variable is	Value of test statistic
tauh0	108.46		
tauh1	246.71		
tauh3	285.51		
tauh5	671.31		
tauh15	3430.43		
taudph0	1016.33		
		taudah0	5.80
		taudah3	4.39
		tausah0	0.00
tausah3	424.85		

The feature to note in the table is that passive learning and active learning seem to yield very different results: while passive learning results in rejection of the null hypothesis of no Granger causality from R to r, active learning seems not to have this implication (except for the static objective function, nondiffuse prior case). It turned out however that this result is a spurious one: autocorrelation of errors in equations (40) and (41) is strong for the equations that led to non-rejection of H_0 (markedly stronger than for the other equations). Note that this finding is based on m-statistics (Kmenta, (1986, p.333)), rather than the Durbin - Watson statistics reported automatically with the regressions, since the latter are inconsistent due to the presence of lagged tau's. As already pointed out the solution in this situation is to include further lags. The autocorrelation properties improved for the case of 8 lags and became acceptable for 12 lags for all equations considered. Results are presented in Table 3 only for the active learning situations, since for the passive learning cases the earlier verdict was repeated.

Table 3.		Exogeneity Test	
Granger Variant		Lag Length: 12	
Reject H_0 when LHS variable is	Value of test statistic	Accept H_0 when LHS variable is	Value of test statistic
taudah0	37.73		
taudah3	163.86		
tausah0	552.50		
tausah3	759.63		

Table 4 displays results of the Sims variant for precisely the same set of situations as Table 2. The results provide unambiguous support for the claim that the theoretical results in chapter III show up in an empirically detectable way in the data:

Table 4.		Exogeneity Test	
Sims Variant		Lead/Lag Length = 5	
Reject H_0 when LHS variable is	Value of test statistic	Accept H_0 when LHS variable is	Value of test statistic
Rh0	139.54		
Rh1	398.96		
Rh3	827.89		
Rh5	1124.45		
Rh15	7584.69		
Rdph0	578.24		
Rdah0	2050.27		
Rdah3	6180.43		
Rsah0	268.34		
Rsah3	598.47		

The interpretation of these test results is the following. Strict exogeneity of r in the presence of learning has been firmly rejected for various situations in the generated data. Thus in empirical work, the investigator can reasonably be assumed to be able to detect effects of this sort. The same applies for strong exogeneity since one of its components, Granger noncausality from R to r was subjected to a direct test and rejected. The other concepts are not directly testable, but indirectly those are also seen to be affected in an empirically detectable way as chapter III predicted.

Notes

1.   I am grateful to Roberto Mariano for pointing out to me that the qualifications appearing in the latter two remarks are necessary.

VI Summary, Directions for Future Research

VI.1 Summary

The study has explored the econometric implications of the presence of an agent in the data generating mechanism who performs learning. It related the issues addressed to major areas in the literature. An illustrative model was presented in which even in the absence of any explicit dynamics the problem became nontrivial and dynamic by virtue of the presence of learning about the unknown parameter. The agent had to strike an optimal balance between current payoff maximization and generation of information in the future. The distinction between active and passive learning was made. On the basis of an argument on the curvature of the value function arising in the problem, active learning was shown to be generically optimal.

Following a description of the model, further definitions needed for precise treatment of the issues were given. Then the issue of exogeneity of policy variables was addressed. It was demonstrated that learning by the policymaker is incompatible with weak, strong and super exogeneity of policy variables in a reasonably specified econometric model. Learning was shown to have implications analogous to the subject matter of the Lucas critique. A simple addition to augment the definitions of exogeneity in Engle, Hendry and Richard (1983) was proposed.

Chapters IV and V reported the results of a simulation exercise based on the model. Active learning was shown to be a distinct cause of time inconsistency of optimal plans, and this phenomenon has been quantitatively characterized. The data generated were then used to perform empirical exogeneity tests for the various exogeneity concepts encountered. These provided evidence that the effects predicted by theory described in chapter III show up in empirically detectable way in the data.

VI.2 Directions for Future Research

VI.2.1 Time Deformation and Learning

The point has been made in chapter III that observed time series arising from optimizing decisions of a learning rational agent will not be stationary. The graphs presented also illustrate this point. We need not stop after having made this observation however. It is a specific kind of nonstationarity that we encountered: the increments in the flow of relevant information available to the agent affect the process generating the data on policy variables we observe. It may turn out that the nature of nonstationarity in the data observed by the econometrician may be conveniently characterized using the notion of subordinated processes. The flow of information - the product of learning - causes decision rules to be revised which in turn implies that decisions will be different - to a larger or lesser degree depending on the increment of relevant information that was generated.

Assuming that Bayesian updating can represent learning and denoting the agent's current beliefs by $p(t)$, it is $\Delta p(t)$ that "directs" the process generating the values we observe. It is only our incorrect view (to the extent that learning is ignored) about the structure of the process generating the data and tradition that makes us index it with calendar time. The correct way to view them may be $R\{\Delta p(t)\}$ instead of $R(t)$ and $r\{\Delta p(t)\}$ instead of $r(t)$.

Subordination is a random sampling problem. Because of random sampling the fundamental process cannot be directly estimated, since noise is coupled with effects of random sampling. Evidently, if a model for the random sampling component is available, the true structure can be inferred. Stock's work on time deformation is along these lines (cf. Stock (1987), and (1988)). He estimated the block generating the increments in information. In our context, the assumption of rationality can be argued to imply Bayesian updating, hence the analogous process need not be estimated. The exact level of information available to the agent is not observable for the outside econometrician however, therefore that needs to be estimated.

In the finance literature on subordinated processes $\Delta p(t)$ is in fact called the directing process, and $R\{p(t)\}$ is said to be subordinate to $R(t)$. A given increment in $p(t)$ will induce a change in the decision rule and thus in the observed R value that is given by a stable relationship. Given a stationary environment and Bayesian updating, if one is willing to make the assumptions of chapter II, $\Delta p(t)$ is quantifiable by the increments in the scalar parameter s_t, as shown in section II.4.

Modelling learning as Bayesian updating of priors is consistent with increments of information being always nonnegative, a requirement on directing processes.

Note that with passive learning the flow of information is

exogenous. This is meant not in the sense that the probability distribution of the information gain is independent from the probability distribution of the elements of the model. Rather it is a consequence of the quite arbitrary way the limits of the model are drawn: information gain is a byproduct of the decision process not optimized for.

Active learning endogenizes the flow of information by explicitly optimizing simultaneously for current reward and the amount of information generated. Thus with active learning the fundamental process generating $r(t)$ and $R(t)$ and the directing process generating $\Delta p(t)$ are contemporaneously correlated. This violates the definition of subordinated processes, therefore a suitable generalization would need to be found.

The above proposed methodology may be capable of handling portions of time series which are nonstationary due to the presence of learning and it simply collapses to the usual calendar-time based methods when (and if) beliefs converged to the truth. Thus a general methodology may be obtainable for time series with learning effects in them. This would be a major feat since econometrics as it is now understood usually starts out by assuming an equilibrium with constant information sets for agents.

VI.2.2. Incomplete Learning on the Long Run

Learning, as discussed in the study was essentially learning about a single aspect of a stable environment. In this setup, the belief distribution always converged to full and precise knowledge of the initially unknown parameter. This seems to imply that the entire literature on learning focusses on a transitory phenomenon. The examples in McLennan (1987) and Feldman (1988-89) indicate that this may be an overly restrictive view on the role of learning in economic models (they give examples for finite and infinite action space respectively, in which beliefs do not converge to the truth with positive probability). Thus the possibility opens up that incomplete learning on the long run in this sense may be a generic outcome of learning algorithms. This may refute the view that learning is merely of transitory importance since in the limit agents necessarily have complete and precise information about the environment they are facing. The possibility of incomplete learning also gives a boost to the weight results of this study carry. For example if a hypothesized regime change possibly implied beliefs converging to a completely different limit, this could totally destroy the validity of all coefficient estimates in the model which involve (directly or indirectly) the decisionmaker whose decisions are based on these beliefs. Similarly, a regime change implying changes in the assumptions on the rate of risk aversion or other parameters having to do with higher moments of the belief distribution would imply a different set of active learning policy variables and possibly vastly different degrees of time inconsistency of optimal plans if they could change the limit to which beliefs converge.

Studying the structure of these examples, showing that for a wide class of priors, discount rates and distributional assumptions incomplete learning on the long run can occur with non-negligible probability is therefore an avenue of research offering substantial potential rewards.

VI.2.3 On Intertemporal Transfer of Resources

In the formulation of the model used in the study the only avenue for intertemporal transfer of resources was indirectly, via accumulating information when actively learning. If the specification of the model involved other instruments for intertemporal resource transfer such as money for example, interesting trade-offs between the two means of transfer would open up. In a general equilibrium framework with perfectly functioning markets the market rate of interest would equal the discount rate but imperfections - note that less than complete information can be such - would imply that this trade-off affects the optimal controls chosen.

VI.2.4 Learning and Chaos

Simultaneity introduced by active learning as described in section VI.2.1 may imply chaotic behavior. Chaos would arise if the relationship $R \underset{\longleftarrow}{\overset{\longrightarrow}{}} \Delta p$ could be formulated in the form of one of the simple bivariate dynamic systems leading to chaotic behavior. If indeed such a system was found, it would constitute another avenue for proving that different paths of controls could lead to dramatically different limit beliefs, because a chaotic system displays extreme sensitivity to initial conditions.

APPENDIX A

This Appendix introduces a simple additive Laffer curve derived from first principles using a single tax rate. It also develops a general sufficient condition for the occurrence of active learning by a policymaker maximizing the discounted sum of tax revenues. The elements in the formulation are total output and amount of output that is concealed (evasion). Both quantities are sensitive to the tax rate. Recorded output is equal to total output $Q(\tau)$ minus evasion, $S(\tau)$:

$$q(\tau) = Q(\tau) - S(\tau).$$

Tax revenue is

$$\bar{R} = \tau \, q(\tau).$$

Assuming that $Q'(\tau) < 0$ and $S'(\tau) > 0$ will ensure $q'(\tau)<0$. Making the further assumption that $S(\tau) < Q(\tau)$ for all $\tau \in [0,1)$ and $S(1) = Q(1)$ is sufficient for the emergence of a Laffer curve as a reduced form:

$$\bar{R}(0) = \bar{R}(1) = 0 \qquad\qquad\qquad (A1)$$

with the curve having a unique maximum at $\tau^* \in (0,1)$.

The next step is to specify concrete functions for $Q(\tau)$ and $S(\tau)$ and obtain the resulting Laffer curve. To keep matters simple, assume that

$$Q(\tau) = \alpha - \beta \tau \quad , \quad \text{where } \alpha > \beta > 0 \text{ are unknown constants.}$$

Denote $Q(1)$ by $\gamma = \alpha - \beta$ and let

$S(\tau) - \gamma \tau^2$ which results in

$q(\tau) - \alpha - \beta \tau - \gamma \tau^2.$

Introduce the new parameter $\theta - \frac{\beta}{\alpha}$ and rewrite:

$q(\tau) - \alpha - \theta \alpha \tau - (\alpha - \theta \alpha) \tau^2$.

Thus

$q(\tau) - \alpha [1 - \theta \tau - (1 - \theta) \tau^2]$,   yielding

$\bar{R}(\tau) - \alpha \tau [1 - \theta \tau - (1 - \theta) \tau^2].$

Since the maximizing value of τ is equivalent for $\bar{R}(\tau)$ and $\bar{R}(\tau)/\alpha$, the policymaker can be viewed as maximizing the latter expression. This reduces the number of parameters in the policymaker's problem to 1, since

$$R(\tau) - \frac{\bar{R}(\tau)}{\alpha} - \tau [1 - \theta \tau - (1 - \theta) \tau^2]. \qquad (A2)$$

To proceed, assume that the policymaker has a finite, known time horizon: $t - 1, 2, \ldots T$. Current period tax revenue depends only on the current τ_t and a random effect (yet to be introduced). Thus the policymaker's experience with the various tax rate levels provides noisy information about the efficacy of particular tax rates in raising tax revenue. The policymaker maximizes the present value of expected discounted tax revenues generated as a result of the sequence of tax rates imposed on the economy over the time horizon, given the information level acquired on the parameter θ.

To make room for learning some randomness must be imposed. We assume that $R(\tau_t)$ is observed with error. Instead of observing the revenue given by (A2), the policymaker observes $R(r_t) + \epsilon^*(r_t)$, where noise term $\epsilon^*(r_t) = r_t (1 - r_t) \epsilon_t$, $t=1,2,\ldots T$. This is consistent with the deterministic Laffer restrictions (A1). From the definition of the Laffer curve (A2) we have

$$\theta = \frac{R(r_t) - r_t + r_t^3}{r_t^3 - r_t^2} . \qquad \text{(A3)}$$

Instead of this however, the policymaker observes a realization θ_t which is contaminated by noise:

$$\theta_t = \frac{R(r_t) + r_t(1-r_t)\epsilon_t - r_t + r_t^3}{r_t^3 - r_t^2} = \theta - \frac{1}{r_t} \epsilon_t ,$$

or, without loss of generality (since ϵ_t is assumed to possess a probability density function symmetric around 0):

$$\theta_t = \theta + \frac{1}{r_t} \epsilon_t. \qquad \text{(A4)}$$

Observation of $R(r_t)$ with noise may not sound to be a very appealing assumption. However, it is equivalent to postulating additive noise on either one or both of $Q(r_t)$ and $S(r_t)$ and then bringing the resulting aggregate additive noise term to the left hand side. Because the probability density function for noise is symmetric around zero, the result is the formulation proposed. Note that in this interpretation the unknown parameter θ is a constant, but its value is masked by the additive noise term pertaining to $R(r_t)$. There are other possible noise structures, e.g. those that are capable of distinguishing reducible and irreducible randomness in the setup. Here the value of θ can be learnt with arbitrarily high precision over time: there is only reducible randomness.

We now have θ an unknown constant, and its noisy realization, θ_t a random variable. Assuming $\epsilon_t \sim N(0,\sigma^2)$ and for simplicity setting $\sigma^2 = 1$, it follows that $\theta_t \sim N(\theta,\tau_t^{-2})$, or if precision is displayed instead of variance, $\theta_t \sim N(\theta,\tau_t^2)$. A conjugate prior probability distribution for θ is $P_0(\theta) = N(\mu,\rho)$. Another choice for the distribution of ϵ_t would alter the set of convenient conjugate priors for θ.

Assuming Bayesian updating (DeGroot, 1970, p. 162), the following result is obtained:

$$P_1(\theta) = N\left(\frac{\rho\mu + \tau_1^2\theta_1}{\rho + \tau_1^2}, \rho + \tau_1^2\right).$$

Proceeding in the same manner given a sample consisting of

$$w_{(t)} = \begin{bmatrix} \theta_1, & \theta_2, & \cdots & \theta_t \\ \tau_1, & \tau_2, & \cdots & \tau_t \end{bmatrix},$$

the latest posterior is obtained as:

$$P_t(\theta) = N\left(\frac{\rho\mu + \sum_{i=1}^{t}\tau_i^2\theta_i}{\rho + \sum_{i=1}^{t}\tau_i^2}, \rho + \sum_{i=1}^{t}\tau_i^2\right). \tag{A5}$$

At the beginning of each period t the policymaker has already observed $w_{(t-1)}$. Given this information and current beliefs $p_t(\theta)$, he chooses τ_t so as to strike an optimal balance between current revenue and future information gains that can yield extra revenue in later periods. The existence of this second element makes an intertemporal trade-off possible.

The next step is to capture this intertemporal aspect by casting the problem in a dynamic programming framework following Easley and Kiefer (1988) and Grossman, Kihlstrom and Mirman (1977). The policymaker maximizes the following expression:

$$
E_0 \left\{ \sum_{t=0}^{T} \delta^t \left\{ R(\tau_t) \mid p_t(\theta) \right\} \right\} - E_0 \left\{ \sum_{t=0}^{T} \delta^t \left\{ E_t \left[R(\tau_t) \mid p_t(\theta) \right] \right\} \right\}, \quad (A6)
$$

where $\delta \in (0,1]$ is a known constant and $p_t(\theta)$, i.e. beliefs, evolve according to Bayes' rule as summarized in (A5).

Note that $E_t(\, . \, \mid p_t(\theta))$ denotes the expectation taken with respect to the latest posterior distribution embodying current beliefs about θ. The period 0 posterior is taken to be the initial prior distribution. Now define the value function as

$$
V^T(p_0(\theta)) = \max_{\{\tau_t\}_{t=0}^{T}} E_0 \left\{ \left\{ R(\tau_0) \mid p_0(\theta)) \right\} + \sum_{i=1}^{T} \delta^t \, E_t \left\{ R(\tau_t) \mid p_t(\theta) \right\} \right\}
$$

or equivalently in the recursive form:

$$
V^T(p_0(\theta)) = \max_{\tau_0 \in [0,1]} E_0 \left\{ \left\{ R(\tau_0) \mid p_0(\theta) \right\} + \delta \, E_1 \left\{ V^{T-1} \left[p_1(\theta) \right] \right\} \right\}. \quad (A7)
$$

The first term on the right-hand side of (A7) represents current revenue, i.e. current gains attainable, whereas the second is expected maximum future revenue given the amount of information generated by choice of the control variable at level τ_0 in the present period and assuming Bayesian updating (cf. Grossman, Kihlstrom and Mirman (1977)). This second term contains the gains that can be obtained by learning. At optimum, the policymaker strikes a balance between present revenue and future gains due to "sharpened" information.

At this point we have at hand a "technology" jointly producing government revenue and information. The input to this "technology" from the point of view of the policymaker in any period is the control variable r_t and the output is current revenue and information. Raw information is contained in the observation pair $w_t - (r_t, R_t)$, where $R_t - R(r_t) + \epsilon^*(r_t)$ and processed information is embodied in the probability density function $p_t(\theta)$. Note that $r_t - 0$ or $r_t - 1$ provides no information at all on θ. Such levels of the control result in $R_t - 0$ irrespective of the value of θ.

Let superscript P stand for passive learning, A for active learning. Define expected total discounted revenue (TDR) with passive learning only from the vantage point of period t as

$$TDR_t^P - E_t \left\{ \sum_{j-t}^{T} \delta^{j-t} R_j \left[r_j^P , p_j^P(\theta) \right] \right\} \qquad t-1,2,\ldots T.$$

The corresponding quantity with active learning is

$$TDR_t^A - E_t \left\{ \sum_{j-t}^{T} \delta^{j-t} R_j \left[r_j^A , p_j^A(\theta) \right] \right\} \qquad t-1,2,\ldots T.$$

A necessary and sufficient condition for experimentation to occur in period t, i.e. for r_t^A to be chosen by the policymaker instead of r_t^P is thus

$$TDR_t^A > TDR_t^P. \qquad (A8)$$

This statement is more than an obvious truism only if it can specified in terms of the parameters of the prior $p_0(\theta)$, δ and the controls to be chosen.

APPENDIX B

This Appendix derives the result referred to in chapter II as the
"bigger is better" result. It states that a larger magnitude of the control
variable leads to increased information gain. Using standard arguments it is
easily shown that given a Bayesian updating procedure on the probability
distribution of the unknown parameters and assuming normality, control
variables of the largest possible magnitude in the first couple of periods
are optimal from the point of view of information generation. The derivation
yields the same result as the one obtained in chapter II with the Laffer
curve notation. The final portion of this Appendix derives (8) and (9), the
Bayesian updating formulae.

Consider the following simple linear control problem with scalar
variables and parameter:

$$y_t = \beta \, r_t - u_t \qquad\qquad\qquad t = 1, 2, \ldots, T$$

where y_t and r_t are the state and control variables, respectively and
$u_t \sim$ iid $N(0, \sigma_u^2)$.

The parameter β is unknown. Prior information on it is assumed to
be embodied in the prior distribution $N(m_0, \sigma_0^2)$, possibly obtained as a
previous estimation result. This distribution is updated in each period.
Suppose for simplicity that σ_0^2 and σ_u^2 are known quantities. Then it is easy
to verify (e.g. Raiffa and Schleifer, 1961, p. 337) that the posterior
probability distribution of the parameter β is $N(m_t, \sigma_t^2)$, where

$$m_t = \frac{\dfrac{m_0}{\sigma_0^2} + \dfrac{1}{\sigma_u^2} \sum\limits_{i=1}^{t} r_i y_i}{\dfrac{1}{\sigma_0^2} + \dfrac{1}{\sigma_u^2} \sum\limits_{i=1}^{t} r_i^2} \quad \text{and} \quad \sigma_t^2 = \frac{1}{\dfrac{1}{\sigma_0^2} + \dfrac{1}{\sigma_u^2} \sum\limits_{i=1}^{t} r_i^2} \; .$$

It is seen that large values of the control variable r_t in the initial periods decrease the variance of all subsequent posterior probability distributions on β. If it is the true parameter value $\tilde{\beta}$ that m_t converges to, this always amounts to an increase in the precision of information obtained on the unknown parameter β. Hence the claim, the larger the better for the initial control variables. (This result is also obvious from (A5)).

Although this result follows formally from the formulation presented, relaxing some of the assumptions or imposing plausible restrictions on the variables can invalidate it. Foremost is the fact that allowing for more than one unknown parameter destroys this property. If imprecision in measurement and linear Gaussian Kalman filtering is assumed, we get the opposite extreme, namely that informational gain in any period is independent of the magnitude of the controls applied prior to that period (Anderson and Moore 1979, p. 41). In a more general Kalman filtering environment we get the intuitive result that the gain in precision is a function of the control employed but it is no longer true that "the bigger the better". The result thus rests on a rather restrictive set of assumptions:

- only one parameter is unknown

- normal prior on the parameters

- sequential Bayesian updating

- noise is i.i.d. (thus feedforward control is ruled out)

- observation error is zero (or at least additive, in a
  linear model)

- no restrictions on the magnitude or variance of the
  control variables, in particular controls are
  costless to apply

- the sequence of posterior means converges to the true
  parameter value

- no intrinsic dynamics in the maximand (as seen in section
  IV.3.1)

Let us now verify (8) and (9), the formulae for Bayesian updating
of the belief distribution's moments. To consolidate notation with that of
chapter II, denote by $h_t - \sigma_t^{-2}$ and without loss of generality let $\sigma_u^2 - 1$.
Then

$$h_t - h_0 + \sum_{i-1}^{t} r_i^2 \quad \text{and}$$

$$h_{t+1} - h_0 + \sum_{i-1}^{t+1} r_i^2 - h_t + r_i^2 \; ,$$

verifying (8). From the above and the formula for m_t we have

$$m_t h_t - \frac{m_0}{\sigma_0^2} + \sum_{i-1}^{t} r_i y_i \; , \quad \text{consequently}$$

$$m_{t+1}h_{t+1} = \frac{m_0}{\sigma_0^2} + \sum_{i=1}^{t} r_i y_i + r_{t+1} y_{t+1}$$

$$= m_t h_t + r_{t+1} y_{t+1}.$$

Replacing y_t by the corresponding quantity in chapter II: e_t, defined by (7), we obtain (9).

APPENDIX C

This Appendix contains two proofs: that of the Lemma in chapter II.2, and of the validity of equation (17). Let us start with the proof of the Lemma.

Lemma: $v_t(p)$, $p \in P$ is convex.

We shall prove it by backward induction for a problem with known, finite horizon, starting from the last period. Since we have $v_{T+1} = 0$, denoting the period T posterior probability distribution by p_T, the final period value function is

$$v_T(p_T) = R(r_T^* | p_T) = \max_{r_T} E_T \ R(r_T) \ .$$

We have

$$v_T(p_T) = \max_{r_T} \int_{-\infty}^{\infty} \alpha \ r_T(1-r_T) - \beta \ r_T^2(1-r_T) \ d\Phi\left(\frac{\beta - m_T}{1}\right) =$$

$$= \max_{r_T} \alpha \ r_T(1-r_T) - r_T^2(1-r_T) \int_{-\infty}^{\infty} \beta \ d\Phi\left(\beta - m_T\right) =$$

$$- \max_{r_T} \alpha \ r_T(1-r_T) - m_T \ r_T^2(1-r_T) \ ,$$

Thus the solution is in terms of the certainty equivalent. The reason is that this is the final period. This value function is now shown to be convex in beliefs. To simplify, let $M^1(r_t) = \alpha \ r_T(1-r_T) - m_T^1 \ r_T^2(1-r_T)$, where m_T^i denotes $E_T(\beta)$ calculated according to the prior belief p_T^i in period T, for i=1,2. Also, let $m_T^\lambda = \lambda \ m_T^1 + (1-\lambda) \ m_T^2$ and $p_T^\lambda = \lambda \ p_T^1 + (1-\lambda) \ p_T^2$ for $\lambda \in [0,1]$. Also note that the normal family of distributions is closed under multiplication by a constant and addition, thus a convex linear combination of two normal distributions is still a normal distribution.

Convexity in beliefs means

$$\lambda v_T(p_T^1) + (1-\lambda)v_T(p_T^2) \geq v_T(p_T^\lambda), \qquad (C1)$$

or

$$\lambda \max_{r_T} \left\{ \alpha r_T(1-r_T) - m_T^1 \ r_T^2(1-r_T) \right\} + (1-\lambda) \max_{r_T} \left\{ \alpha r_T(1-r_T) - m_T^2 \ r_T^2(1-r_T) \right\}$$

$$\geq \max_{r_T} \left\{ \alpha \ r_T(1-r_T) - m_T^\lambda \ r_T^2(1-r_T) \right\} \ .$$

Note that except in the case of r_t, superscripts are not exponents. For λ equal to 0 or 1, we have linearity. For $\lambda \in (0,1)$ rewrite the above inequality in a somewhat simpler form, using the notation introduced above, as

$$\lambda \max_{r_T} M^1(r_T) + (1-\lambda) \max_{r_T} M^2(r_T) \geq \max_{r_T} \left\{ \lambda \ M^1(r_T) + (1-\lambda) \ M^2(r_T) \right\}.$$

This is transformed until a form that is known to be true is reached:

$$\frac{\lambda}{1-\lambda} \max_{\tau_T} M^1(\tau_T) + \max_{\tau_T} M^2(\tau_T) \geq \frac{1}{1-\lambda} \max_{\tau_T} \left\{ \lambda \, M^1(\tau_T) + (1-\lambda) \, M^2(\tau_T) \right\}$$

$$\max_{\tau_T} \frac{\lambda}{1-\lambda} M^1(\tau_T) + \max_{\tau_T} M^2(\tau_T) \geq \max_{\tau_T} \left\{ \frac{\lambda}{1-\lambda} M^1(\tau_T) + M^2(\tau_T) \right\}$$

$$\max_{\tau_T} \tilde{M}^1(\tau_T) + \max_{\tau_T} M^2(\tau_T) \geq \max_{\tau_T} \left\{ \tilde{M}^1(\tau_T) + M^2(\tau_T) \right\} .$$

The last line is evident for $\max_{\tau_T} \tilde{M}_1(\tau_T)$ and $\max_{\tau_T} M_2(\tau_T)$ which exist and are nonnegative. Unique existence is proven for the one period problem in chapter II and nonnegativity is evident from the specification. Thus we have the first step of the proof: the final period value function $v_T(p_T)$ is convex in beliefs. It also follows (from the linearity of the conditional expectation operator) that $E_{T-1}v_T(p_T)$ is convex in beliefs. Now we proceed by proving that $v_{T-1}(p_{T-1})$ is also convex, given that $E_{T-1}v_T(p_T)$ is convex:

$$\lambda v_{T-1}(p^1_{T-1}) + (1-\lambda)v_{T-1}(p^2_{T-1}) \geq v_{T-1}(p^\lambda_{T-1}). \qquad (C2)$$

To prove this, we start out by the left hand side of (C2), consider transformations that decrease its value (or at most leave it unchanged) and finally end up with the right hand side quantity. Let p^1_T be the posterior belief distribution in period T arising via Bayesian updating if the prior was p^i_{T-1}, $i=1,2$. We have in (C2):

$$\lambda v_{T-1}(p^1_{T-1}) + (1-\lambda)v_{T-1}(p^2_{T-1}) = \lambda \max_{\tau_{T-1}} E_{T-1}\left[M^1(\tau_{T-1}) + \delta \, v_T(p^1_T) \right] +$$

$$+ (1-\lambda) \max_{\tau_{T-1}} E_{T-1}\left[M^2(\tau_{T-1}) + \delta \, v_T(p^2_T) \right] \geq$$

$$\geq \max_{\tau_{T-1}} E_{T-1}\left\{ \lambda \left[M^1(\tau_{T-1})+\delta v_T(p^1_T) \right] + (1-\lambda)\left[M^2(\tau_{T-1})+\delta v_T(p^2_T) \right] \right\}$$

$$- \max_{r_{T-1}} E_{T-1} \left\{ M^{\lambda}(r_{T-1}) + \delta \left[\lambda v_T(p_T^1) + (1-\lambda) v_T(p_T^2) \right] \right\} \geq$$

$$\geq \max_{r_{T-1}} E_{T-1} \left\{ M^{\lambda}(r_{T-1}) + \delta v_T(p_T^{\lambda}) \right\} - v_{T-1}(p_{T-1}^{\lambda}) .$$

The first inequality follows from the properties of the maximum operator and the second one follows from (C1). This completes the proof since (C2) is implied. It is useful to point out (as C. Gourieroux did) that the proof goes through without having to use the explicit form of the maximand. Hence the Lemma holds for a more general class of problems including the one considered in this study.

The second part of this Appendix contains proof of the validity of equation (17):

$$V_t \left(km_t , \frac{h_t}{k^2} \right) = V_t(m_t , h_t) \qquad \text{for all } k \neq 0. \qquad (17)$$

Clearly $\alpha - \beta r_t + u_t = \alpha - (k\beta)(\frac{r_t}{k}) + u_t$ for all $k \neq 0$, and the assumption of the prior $N(m_t, h_t)$ for β implies $N(km_t, \frac{h_t}{k^2})$ as prior for $k\beta$. (Remembering that h_t denotes precision, the reciprocal of variance, this is also obvious.) We now have that the part in the value function containing current quantities is identical for the two different beliefs specified on the two sides of equation (17). To complete the proof it needs to be proven that the part containing future quantities is also identical for the two cases. Let

$$B \begin{pmatrix} m_t \\ h_t \end{pmatrix} = \begin{pmatrix} m_{t+1} \\ h_{t+1} \end{pmatrix}$$

represent the update formula defined by (8) and (9). We seek to show that

$$V_{t+1}\left(B\begin{pmatrix}m_t\\h_t\end{pmatrix}\right) = V_{t+1}\left(B\begin{pmatrix}m_t\\\dfrac{h_t}{k^2}\end{pmatrix}\right)$$

To do so we systematically check the ingredients of the problem under the two alternatives. Let one prime denote quantities in the case when beliefs are as on the right hand side of (17) and double prime the case when beliefs are as on the left hand side of (17). Note that it follows from the discussion on the current quantities that the optimal tax rate applied in the latter case will be $\dfrac{r_t}{k}$. For the first case we obtain utilizing (8) and (9) that

$$h'_{t+1} = h'_t + r_t^2 \qquad \text{and}$$

$$m'_{t+1} = \frac{m'_t h'_t + r_t(\beta r_t - u_t)}{h'_{t+1}}.$$

For the second case we get:

$$h''_{t+1} = \frac{h'_t}{k^2} + \left(\frac{r_t}{k}\right)^2 = \frac{h'_{t+1}}{k^2}.$$

Also

$$m''_{t+1} h''_{t+1} = km'_t \frac{h'_t}{k^2} + \frac{r_t}{k}\left(k\beta \frac{r_t}{k} - u_t\right) = \frac{1}{k}\left[m'_t h'_t + r_t(\beta r_t - u_t)\right].$$

The preceding two equations directly imply

$$m''_{t+1} = km'_{t+1}.$$

This completes the proof: one period later the same two cases arise, again yielding identical "current" solutions and so on until the final period when there is only the "current" period.

APPENDIX D

This Appendix contains the derivation of the first and second partial derivatives of s_{t+1} with respect to r_t. The expectation at time t of both is shown to be positive. Then a probabilistic sufficient condition is given which ensures that the problem is well defined in each period. This result formalizes the argument of note 1 in chapter II. An argument follows that with a positive prior mean of the belief distribution, the mean belief in later periods is also positive, thus completing the proof that the problem under scrutiny is a meaningful one. Throughout the Appendix it will be a maintained assumption that $m_t > 0$ in each period. The circumstances under which $m_t \leq 0$ would occur are examined at the end of this Appendix.

From the definition of s_t and equations (21), (22) in section II.3, we have

$$s_{t+1} = \frac{m_t h_t + r_t e_t}{(h_t + r_t^2)^{1/2}} = \frac{m_t h_t + \beta r_t^2 - r_t u_t}{(h_t + r_t^2)^{1/2}} . \tag{D1}$$

Therefore

$$\frac{\partial s_{t+1}}{\partial r_t} = \frac{(2\beta r_t - u_t)(h_t + r_t^2)^{1/2} - (h_t + r_t^2)^{-1/2} r_t [m_t h_t + \beta r_t^2 - r_t u_t]}{h_t + r_t^2} \tag{D2}$$

Now take expectations with respect to the random variable u_t and condition on the latest information on β:

$$\frac{\partial s_{t+1}}{\partial r_t} = 2 m_t r_t (h_t + r_t^2)^{-1/2} - (m_t h_t r_t - m_t r_t^3)(h_t + r_t^2)^{-3/2} .$$

Multiplying by the positive quantity $(h_t + r_t^2)^{3/2}$ does not change the sign of our partial derivative. We get:

$$2m_t h_t r_t + 2m_t r_t^3 - m_t h_t r_t - m_t r_t^3$$

$$- m_t h_t r_t + m_t r_t^3 > 0.$$

Therefore the first partial derivative is positive: we encountered the bigger is better result again. Now we need to check if this property carries over to the more general case of s_j, $j > 2$ as well. Take $j = t + 2$:

$$s_{t+2} = \frac{m_t h_t + r_t e_t + r_{t+1} e_{t+1}}{(h_t + r_t^2 + r_{t+1}^2)^{1/2}} - \frac{m_t h_t + \beta r_t^2 - r_t u_t + \beta r_{t+1}^2 - r_{t+1} u_{t+1}}{(h_t + r_t^2 + r_{t+1}^2)^{1/2}}$$

Note that the numerator is necessarily positive. Also, by (7), s_{t+j} contains r_t only in the parts already included in s_{t+1}. This is sufficient to make $\frac{\partial s_{t+j}}{\partial r_t} > 0$ as well. Finally, it is argued that partials with respect to r_{t+i}, $t < i < j$ need not be considered since only r_t is actually applied in period t: planned optimal tax rates for the future can and do change as time passes and new information becomes available. This phenomenon, which is a form of time inconsistency is discussed in detail in section IV.3. An argument similar to this one applies for the case $j = t+3$, etc.

Let us now turn to the second partial derivative. Partially differentiating (D2) with respect to r_t yields

$$\frac{\partial^2 s_{t+1}}{\partial r_t^2} = \frac{\partial}{\partial r_t}\left[\frac{2\beta r_t - u_t}{(h_t + r_t^2)^{1/2}} - \frac{r_t m_t h_t + \beta r_t^3 - r_t^2 u_t]}{(h_t + r_t^2)^{3/2}}\right]$$

$$- \frac{2\beta(h_t+r_t^2)^{1/2} - r_t(h_t+r_t^2)^{-1/2}[2\beta r_t-u_t]}{h_t+r_t^2} -$$

$$- \frac{[m_t h_t + 3\beta r_t^2 - 2r_t u_t](h_t+r_t^2)^{3/2} - 3r_t[m_t h_t r_t+\beta r_t^3-r_t^2 u_t](h_t+r_t^2)^{1/2}}{(h_t+r_t^2)^3}$$

Now take expectations with respect to the random variable u_t and condition on the latest information on β, then multiply with the positive quantity $(h_t+r_t^2)^{5/2}$ to obtain

$$2m_t(h_t+r_t^2)^2 - 2m_t r_t^2(h_t+r_t^2) - m_t h_t(h_t+r_t^2) - 3m_t r_t^2(h_t+r_t^2) +$$

$$+ 3m_t h_t r_t^2 + 3m_t r_t^4 -$$

$$- m_t h_t(h_t + r_t^2) - m_t h_t h_{t+1} > 0.$$

Therefore the time t expectation of the second partial is also positive: there are increasing returns (in terms of precision) to increasing the magnitude of the control variable in order to gain more precise information in the next period. Note that since the "cost" term: the value of foregone payoffs is not accounted for, this does not necessarily imply increasing returns to experimentation by the policymaker.

Let us now turn to obtaining an overall necessary and sufficient condition for the problem of the policymaker to be well defined. It will merge three conditions: one is (11) from section II.1, rewritten as

$$\alpha - m_t r_t > m_t[2r_t+1] . \tag{D3}$$

The other condition ensures that the probability of a realization of the random noise component u_t which makes government revenue negative is smaller than a prespecified level: ω. There is an alternative way to achieve this: to truncate the support of the distribution of the noise term to ensure that it cannot occur. Since this would not necessarily be a minor truncation however, this approach is not preferred.

The event $R(r_t) < 0$ is equivalent to $u_t < m_t r_t - \alpha$. Since u_t is distributed as $N(0, s^2)$, the probability of the event that government revenue is negative in any period is equal to

$$\Phi\left(\frac{m_t r_t - \alpha}{s}\right) .$$

Suppose we consider only the case when $m_t r_t - \alpha < 0$. (It is implied by the restriction (D5) to be introduced momentarily). Letting the probability of government revenue being negative in period t to be less than or equal to ω, we get

$$\Phi\left(\frac{m_t r_t - \alpha}{s}\right) = 1 - \Phi\left(\frac{\alpha - m_t r_t}{s}\right) \leq \omega.$$

This implies

$$\Phi\left(\frac{\alpha - m_t r_t}{s}\right) \geq 1 - \omega$$

and thus

$$\alpha - m_t r_t > s \; \Phi^{-1}(1-\omega). \tag{D4}$$

Thus for any given ω the smaller s and the larger α is, the more likely it is that government revenue is nonnegative. For example demanding α to be large relative to the mean belief on β and demanding s to be small amounts to placing an upper bound on the extent uncertainty can affect the payoff of the learning agent. This result seems somewhat counterintuitive but can be explained along the lines of MacRae (1972). Her argument has been discussed in chapter II. The need for restricting α and s by (D4) arises essentially because the support of the belief distribution was not assumed to be compact (cf. assumption (iii) in section II.1). It is the price paid for not making this assumption. An alternative, perhaps more elegant way to achieve nonnegativity is to impose Sargent's projection operator:

$$m_t - m_t^{max} \qquad \text{if } m_t > m_t^{max}$$

where m_t^{max} is the maximum value of the mean of the belief distribution given the current realization of u_t that is compatible with $R(r_t)$ being nonnegative.

The third requirement is the obvious one of restricting the optimal tax rate to be in the [0,1] interval. From (12), nonnegativity of the optimal tax rate is evident, requiring it to be less than 1 boils down to requiring that

$$\alpha > m_t , \qquad (D5)$$

which is implied by (11), or equivalently: (D3).

Now combine (D3) and (D4) using the fact that they have the same structure:

$$\alpha - m_t r_t > \text{Max} \left\{ s \; \Phi^{-1}(1-\omega), \; m_t[2r_t+1] \right\} . \qquad (D6)$$

This condition (together with the maintained assumption $m_t > 0$ for all t) is necessary and sufficient to ensure that the problem is well defined for any period, government revenue is not negative with a pre-specified probability and there is room for active learning by increasing the magnitude of controls applied.

Finally focus on the sign of m_t, the mean of the belief distribution in period t. Suppose that m_1 is positive. The sign of m_t is the same as that of $m_t h_t$, which is simpler to analyse. It is positive if:

$$m_t h_t = m_1 h_1 + \sum_{i=1}^{t-1} r_i e_i = m_1 h_1 + \sum_{i=1}^{t-1} r_i \left[\beta r_i - u_i \right] > 0 \text{ , or}$$

$$m_1 h_1 + \beta \sum_{i=1}^{t-1} r_i^2 > \sum_{i=1}^{t-1} r_i u_i \text{ .}$$

From this it is clear that for $\beta > 0$, $m_t < 0$ occurs only in the very unlikely event that all the following occur:

the prior is close to diffuse;

β itself is very small;

large realizations of u_i occur just when r_i are large;

we are in an early period: the sum of squared r_i's is small.

In the simulations it was possible to generate $m_t < 0$, but only if the constellation of parameters was specifically geared towards achieving this goal. Very many repetitions were necessary with a very small α and β and very large variance of the noise variable for a single occurrence of $m_t < 0$ for some (always an "early") t. It was this phenomenon that made the double truncation of the support of the distribution of u_t necessary: the probability that $m_t < 0$ given that $m_1 > 0$ can be made arbitrarily small by choosing a suitable K to define the support of the u as $[-K, K]$. Given the truncation, a sequence of outlier realizations of u close to -K would be necessary to drive m_t negative for some t. The probability of the occurrence of such a sequence is a positive integer power of an already very low probability, therefore it is negligible.

An alternative approach would have been to maintain the assumption of infinite support for the noise term but incorporate a projection operator "escape clause" into the update rule (9): if m_t < 0 as computed by (9), then reset m_t - M, where M is a positive constant. This would rule out the possibility of mean beliefs becoming negative. Note that the selection of reasonable values for the parameters of the model alone would only make the probability of the occurrence of m_t < 0 small, but not zero, since in that case the realization of u_t could be smaller than any fixed lower bound implied by the setting of the values of the parameters. Also, problems arise with this approach, therefore it is only mentioned as a possibility. The problems are the following. First, presumably some resetting rule for the precision would have to be found for the case when the projection is actually applied. Second, the theoretical possibility arises of entering an infinite loop endlessly repeating the projection and hence ruling out convergence. It seems that both problems are surmountable, but this is not pursued here.

Bibliography

Amemiya, T. (1985): Advanced Econometrics

   Cambridge: Harvard University Press

Anderson, B. D. O., J. B. Moore (1979): Optimal Filtering

   Englewood Cliffs: Prentice Hall

Arrow, K. J. (1965): Aspects of the Theory of Risk-Bearing

   Helsinki: Yrjö Jahnsson Lectures

Arrow, K. J. (1978): Risk Allocation and Information: Some Recent
Theoretical Developments,

   First Annual Lecture of the Geneva Association

   (Geneva: Association Internationale pour l'Etude de l'Economie de

   l'Assurance)

Bar-Shalom, Y. E. Tse (1976): Caution, Probing and the Value of Information
in the Control of Uncertain Systems

   Annals of Economic and Social Measurements Vol 5: p323-338

Basmann, R. L. (1965): A Note on the Statistical Testability of "Explicit
Causal Chains" Against the Class of "Interdependent Models"

   Journal of the American Statistical Association Vol 60: p1080-1093

Bellmann, R. (1961): Adaptive Control Processes: A Guided Tour

   Princeton: Princeton University Press

Bertsekas, D. P. (1976): Dynamic Programming and Stochastic Control

   New York: Academic Press

Blume, L. E., M. M. Bray, D. Easley (1982): Introduction to the Stability of Rational Expectations Equilibria

 Journal of Economic Theory Vol 26: p313-317

Bray, M. M., D. M. Kreps (1986): Rational Learning and Rational Expectations

 in: Heller, W. P. et al, editors: Equilibrium Analysis - Essays in

 honor of K. J. Arrow

 Cambridge: Cambridge University Press

Bray, M. M., N. E. Savin (1986): Rational Expectations Equilibria, Learning and Model Specification

 Econometrica Vol 54: p1129-1160

Brunner, K., A. H. Meltzer (1979): Three Aspects of Policy and Policymaking

 New York: North Holland

Calvo, G. A. (1978): On the Time Consistency of Optimal Policy in a Monetary Economy

 Econometrica Vol 46: p1412-1428

Chow, G. (1960): Tests of Equality between Sets of Coefficients in Two Linear Regressions

 Econometrica Vol 28: p591-605

Chow, G. (1981): Econometric Analysis by Control Methods

 New York: John Wiley

Ciccolo, J. (1978): Money, Equity Values and Income - Tests for Exogeneity

 Journal of Money, Credit and Banking Vol 10: p46-64

Crawford, R. G. (1973): Implications of Learning for Economic Models of Uncertainty

 International Economic Review Vol 14: p587-600

Cyert, R. M., M. H. DeGroot (1974): Rational Expectations and Bayesian Analysis

    Journal of Political Economy Vol 82: p521-536

DeCanio, S. J. (1979): Rational Expectations and Learning from Experience

    Quarterly Journal of Economics Vol 92: p47-57

DeGroot, M. H. (1970): Optimal Statistical Decisions

    New York: McGraw - Hill

Dennis, J. E., R. B. Schnabel (1983): Numerical Methods for Unconstrained Optimization and Nonlinear Equations

    Englewood Cliffs, NJ: Prentice-Hall

Easley, D. and N. M. Kiefer (1988): Controlling a Stochastic Process with Unknown Parameters

    Econometrica, Vol 56: p1045-1064

Engle, R. F. (1984): Wald, Likelihood Ratio and Lagrange Multiplier Tests in Econometrics

    Chapter 13 of Griliches, Z., M. Intriligator, eds (1984)

Engle, R. F., D. F. Hendry, J-F. Richard (1983): Exogeneity

    Econometrica, Vol 51: p277-304

Feldman, M. (1988-89): Comment on Kiefer (1988-89)

    Econometric Reviews Vol7: p149-154

Fourgeaud, C., C. Gourieroux, J. Pradel (1986): Learning Procedures and Convergence to Rationality

    Econometrica Vol 54: p845-868

Friedman, B. M. (1979): Optimal Expectations and the Extreme Informational Assumptions of Rational Expectations Macromodels

    Journal of Monetary Economics Vol 5: p23-41

Geweke, J. (1979): Testing the Exogeneity Specification in the Complete Dynamic Simultaneous Equation Model

Journal of Econometrics Vol 7: p163-185

Geweke, J. (1984): Inference and Causality in Economic Time Series

Chapter 19 of Griliches, Z., M. Intriligator, eds (1984)

Geweke, J. (1985): Macroeconometric Modeling and the Theory of the Representative Agent

American Economic Review Vol 75: p206-210

Granger, C. W. J. (1980): Testing for Causality - A Personal Viewpoint

Journal of Economic Dynamics and Control Vol 2: p329-352

Griliches, Z., M. Intriligator, eds (1984): Handbook of Econometrics

New York: North Holland

Grossman, S. J., R. E. Kihlstrom, L. J. Mirman (1977): A Bayesian Approach to the Production of Information and Learning by Doing

Review of Economic Studies Vol 44: p533-547

Grumberg, E., F. Modigliani (1954): The Predictability of Social Events

Journal of Political Economy, Vol162: p465-478

Hansen, L. P., T. J. Sargent (1980): Formulating and Estimating Dynamic Linear Rational Expectations Models

Journal of Economic Dynamics and Control, Vol2

Hausman, J. A. (1978): Specification Tests in Econometrics

Econometrica Vol46: p1251-1271

Hendry, D. F., J-F. Richard (1983): The Econometric Analysis of Economic Time Series

International Statistical Review Vol 51: p11-163

Holly, S., A. Hughes-Hallett (1989): Optimal Control, Expectations and Uncertainty

    Cambridge: Cambridge University Press

Hosoya, Y. (1977): On the Granger Condition for Non-Causality

    Econometrica, Vol 45: p1735-1736

Hughes-Hallett, A., H. Rees (1983): Quantitative Economic Policies and Interactive Planning

    Cambridge: Cambridge University Press

Jovanovic, B., S. Lach (1989): Entry, Exit and Diffusion with Learning by Doing

    American Economic Review: Vol79: p690-699

Kamien, M. F., N. L. Schwartz (1983): Conjectural Variations

    Canadian Journal of Economics Vol16: p191-211

Kendrick, D. A. (1982): Adaptive Control of Macroeconomic Models - Caution and Probing in a Macroeconomic Model

    Journal of Economic Dynamics and Control Vol4: p149-170

Kiefer, N. M. (1988-89): Optimal Collection of Information by Partially Informed Agents

    Econometric Reviews Vol7: p133-148

Koopmans, T. C., ed.(1950): Statistical Inference in Dynamic Economic Models

    Cowles Commission Monograph # 10., New York: John Wiley

Le Cam , L. M., R. A. Olshen, eds (1985): Proceedings of the Berkeley Conference in Honor of J. Neyman and J. Kiefer, June 1983 , Vol 2.

    Monterey: Wadsworth Advanced Books

Lucas, R. E. (1976): Econometric Policy Evaluation: a Critique

in: The Phillips Curve and Labor Markets, Brunner, K. and A. H. Meltzer, eds: p19-46

Carnegie-Rochester Conference Series #1 New York: North Holland

MaCrae, E. C. (1972): Linear Decision with Experimentation

Annals of Economic and Social Measurements Vol 1: p437-447

Marquez, J., P. Pauly (1986): Bayesian Oil Pricing

Revision of Paper at 9th IFAC Conference in Budapest, 1984

Fed, Washington D.C.

McLennan, A. (1987): Incomplete Learning in a Repeated Statistical Decision Problem

Department of Economics, University of Minnesota Working Paper

Mirman, L. J., L. Samuelson, A. Urbano (1989): Monopoly Experimentation

mimeo, Department of Economics, University of Virginia, Charlottesville

Mirman, L. J., L. Samuelson, E. E. Schlee (1990): Strategic Information Manipulation in Duopolies

mimeo, Department of Economics, University of Virginia, Charlottesville

Mizrach, B. (1989): Non-Convergence to Rational Expectations and Optimal Monetary Policy in Models with Learning

Manuscript, Department of Economics, Boston College

Newbold, P. (1978): Feedback Induced by Measurement Errors

International Economic Review Vol 19: p787-791

Pesaran, M. H. (1987): The Limits to Rational Expectations

New York: Basil Blackwell

Prescott, E. C. (1972): The Multiperiod Control Problem Under Uncertainty

Econometrica Vol 40: p1043-1058

Raiffa, H. and R. Schlaifer (1961): Applied Statistical Decision Theory

    Cambridge: MIT Press

Rob, R. (1988): Learning and Capacity Expansion in a New Market under Uncertainty

    Manuscript, University of Pennsylvania, Department of Economics

Rust, J. (1988-89): Comment on Kiefer (1988-89)

    Econometric Reviews Vol7: p155-160

Sargent, T. J. (1981): Interpreting Economic Time Series

    Journal of Political Economy Vol 89: p213-248

Sargent, T. J. (1987): Macroeconomic Theory, Second Edition

    New York: Academic Press

Sargent, T. J., A. Marcet (1987a): Convergence of Least Squares Learning Mechanisms in Self Referential Linear Stochastic Models

    Mimeo, Carnegie-Mellon University and Hoover Institute, Stanford University

Sargent, T. J., A. Marcet (1987b): Convergence of Least Squares Learning in Environments with Hidden State Variables and Private Information

    Mimeo, Carnegie-Mellon University and Hoover Institute, Stanford University

Sims, C. A. (1972): Money, Income and Causality

    American Economic Review Vol 62: p540-552

Sims, C. A. (1972): Are There Exogenous Variables in Short-Run Production Relations?

    Annals of Economic and Social Measurements Vol 1: p17-36

Sims, C. A. (1974): Optimal Stable Policies for Unstable Instruments

    Annals of Economic and Social Measurements Vol3: p257-265

Sims, C. A., (1977): Exogeneity and Causal Ordering in Macroeconomic Models

in: Sims, C. A., ed (1977): New Methods in Business Cycle Research

Proceedings of a Conference in November 1975

Minneapolis: FED of Minneapolis

Siviero, S. (1989): Conjectural Variations Solutions for Static and Dynamic Games

Unpublished Manuscript, University of Pennsylvania, Department of Economics

Spear, S. E. (1989): Learning Rational Expectations under Computability Constraints

Econometrica Vol 57: p889-910

Stock, J. H. (1987): Measuring Business Cycle Time

Journal of Political Economy Vol 95: p1240-1261

Stock, J. H. (1988): Estimating Continuous - Time Processes Subject to Time Deformation

Journal of the American Statistical Association Vol 83: p77-85

Taylor, J. B. (1972): Asymptotic Properties of Multiperiod Control Rules in a Linear Regression Model

Technical Report #79, Economics Series, Stanford University

Theil, H. (1964): Optimal Decision Rules for Government and Industry

Amsterdam: North Holland

Townsend, R. M. (1978): Market Anticipations, Rational Expectations and Bayesian Analysis

International Economic Review Vol 19: p481-494

Townsend, R. M. (1983): Forecasting the Forecasts of Others

Journal of Political Economy Vol 91: p546-588

Tse, E. (1974): Adaptive Dual Control Methods

    Annals of Economic and Social Measurements Vol3: p65-83

Wallis, K. F. (1980): Econometric Implications of the Rational Expectations
Hypothesis

    Econometrica Vol48, #1

Williams, D., C. A. E. Goodhart and D. H. Gowland (1976): Money, Income and
Causality: The UK Experience

    American Economic Review Vol 66: p417-423

Zellner, A. (1988): Optimal Information Processing and Bayes' Theorem

    MRG Working Paper #M8803

    Department of Economics, University of Southern California